Lecture Notes
in Business Information Processing 329

Series Editors

Wil van der Aalst
RWTH Aachen University, Aachen, Germany
John Mylopoulos
University of Trento, Trento, Italy
Michael Rosemann
Queensland University of Technology, Brisbane, QLD, Australia
Michael J. Shaw
University of Illinois, Urbana-Champaign, IL, USA
Clemens Szyperski
Microsoft Research, Redmond, WA, USA

W0080336

More information about this series at http://www.springer.com/series/7911

Mathias Weske · Marco Montali
Ingo Weber · Jan vom Brocke (Eds.)

Business Process Management Forum

BPM Forum 2018
Sydney, NSW, Australia, September 9–14, 2018
Proceedings

 Springer

Editors
Mathias Weske
Hasso-Plattner Institute
University of Potsdam
Potsdam
Germany

Marco Montali ⓘ
Free University of Bozen-Bolzano
Bolzano
Italy

Ingo Weber ⓘ
Data61
CSIRO
Eveleigh, NSW
Australia

Jan vom Brocke
University of Liechtenstein
Vaduz
Liechtenstein

ISSN 1865-1348   ISSN 1865-1356   (electronic)
Lecture Notes in Business Information Processing
ISBN 978-3-319-98650-0   ISBN 978-3-319-98651-7   (eBook)
https://doi.org/10.1007/978-3-319-98651-7

Library of Congress Control Number: 2018950647

This Springer imprint is published by the registered company Springer Nature Switzerland AG
The registered company address is: Gewerbestrasse 11, 6330 Cham, Switzerland

Preface

The 16th International Conference on Business Process Management provided a forum for researchers and practitioners in the broad and diverse field of business process management. To accommodate for the diversity of the field, this year the BPM conference introduced a track structure, with tracks for foundations, engineering, and management. The conference was held in Sydney, Australia, during September 9–14, 2018.

Since its introduction two years ago, the aim of the BPM Forum has been to host innovative research that has high potential of stimulating discussion, but does not quite meet the rigorous quality criteria for the main conference. The papers selected for the forum showcase fresh ideas from exciting and emerging topics in business process management.

In all tracks combined, the reviewing process involved 30 senior Program Committee (PC) members and 99 regular PC members. Each paper was reviewed by a team comprising a senior PC member and a set of regular PC members. Based on the outcome of the reviewing process, we invited 16 innovative papers to the BPM 2018 Forum, out of which 14 followed our invitation. The papers invited to the forum come from a set of 113 submissions that could not be accepted at the main conference.

The BPM 2018 Forum papers in this volume cover topics from new aspects in process modeling, to innovative application areas of BPM, such as the Internet of Things, to managerial aspects including governance and standardization as they pertain to BPM.

We thank the colleagues involved in the organization of the conference, especially the members of the PCs and the Organizing Committee. The development of the program structure was challenging, because with the new track structure this year many more papers were accepted than traditionally at BPM. Still, we managed to provide a packed, exciting program, including the presentations for the 14 Forum papers from this volume.

We thank the conference partner Data61, the Platinum sponsor Signavio, the Gold sponsors Celonis and IBM Research, the Bronze sponsors Bizagi and Springer for their generous support of BPM 2018. We also thank the University of New South Wales and Macquarie University for their enormous and high-quality support.

September 2018

Mathias Weske
Marco Montali
Ingo Weber
Jan vom Brocke

Organization

The BPM Forum was a sub-track of BPM 2018, which was organized by the University of New South Wales, in collaboration with Macquarie University, the University of Technology Sydney, and the Service Science Society, and took place in Sydney, Australia.

Steering Committee

Mathias Weske (Chair)	University of Potsdam, Germany
Boualem Benatallah	University of New South Wales, Australia
Jörg Desel	University of Hagen, Germany
Schahram Dustdar	Vienna University of Technology, Austria
Marlon Dumas	University of Tartu, Estonia
Wil van der Aalst	RWTH Aachen University, Germany
Michael zur Muehlen	Stevens Institute of Technology, USA
Stefanie Rinderle-Ma	University of Vienna, Austria
Barbara Weber	Technical University of Denmark, Denmark
Manfred Reichert	Ulm University, Germany
Jan Mendling	Vienna University of Economics and Business, Austria

Executive Committee

General Chairs

Boualem Benatallah	University of New South Wales, Australia
Jian Yang	Macquarie University, Australia

Program Chairs

Mathias Weske (Consolidation Chair)	University of Potsdam, Germany
Marco Montali (Chair Track I)	University of Bolzano, Italy
Ingo Weber (Chair Track II)	Data61|CSIRO, Australia
Jan vom Brocke (Chair Track III)	University of Liechtenstein, Liechtenstein

Industry Chairs

Fabio Casati	University of Trento, Italy
Gero Decker	Signavio, Germany
Surya Nepal	Data61|CSIRO, Australia

Workshops

Florian Daniel   Politecnico di Milano, Italy
Hamid Motahari   IBM, Almaden Research Center, San Jose, USA
Michael Sheng   Macquarie University, Australia

Demo Chairs

Raffaele Conforti   The University of Melbourne, Australia
Massimiliano de Leoni   Eindhoven University of Technology, The Netherlands
Barbara Weber   Technical University of Denmark, Denmark

Publicity Chairs

Cinzia Cappiello   Politecnico di Milano, Italy
Daniela Grigori   Université Paris Dauphine, France
Oktay Türetken   Eindhoven University of Technology, The Netherlands
Lijie Wen   Tsinghua University, China

Sponsorship and Community Liaison Chairs

François Charoy   University of Lorraine, France
Onur Demirors   Izmir Institute of Technology, Turkey
                              and UNSW Sydney, Australia
Fethi Rabhi   UNSW Sydney, Australia
Daniel Schlagwein   UNSW Sydney, Australia

Panel Chairs

Athman Bouguettaya   The University of Sydney, Australia
Mohand-Saïd Hacid   Université Claude Bernard Lyon 1, France
Manfred Reichert   Ulm University, Germany

Tutorial Chairs

Marcello La Rosa   The University of Melbourne, Australia
Stefanie Rinderle-Ma   University of Vienna, Austria
Farouk Toumani   Blaise Pascale University, France

Doctoral Consortium Chairs

Yan Wang   Macquarie University, Australia
Josep Carmona   Universitat Politècnica de Catalunya, Spain

Mini-Sabbatical Program Chairs

Shazia Sadiq   The University of Queensland, Australia
Moe Thandar Wynn   Queensland University of Technology, Australia

Local Organization Liaison and Networking Chair

Ghassan Beydoun   University of Technology Sydney, Australia

Local Arrangements Committee

Olivera Marjanovic   University of Technology Sydney, Australia
  (Co-chair)
Lina Yao (Co-chair)   UNSW Sydney, Australia
Kyeong Kang   University of Technology Sydney, Australia
Wei Zhang   Macquarie University, Australia

Proceedings Chair

Luise Pufahl   University of Potsdam, Germany

Web and Social Media Chair

Amin Beheshti   Macquarie University, Australia

Track I (Foundations)

Senior Program Committee

Florian Daniel   Politecnico di Milano, Italy
Dirk Fahland   Eindhoven University of Technology, The Netherlands
Giancarlo Guizzardi   Free University of Bozen-Bolzano, Italy
Thomas Hildebrandt   IT University of Copenhagen, Denmark
Marcello La Rosa   The University of Melbourne, Australia
John Mylopoulos   University of Toronto, Canada
Manfred Reichert   Ulm University, Germany
Jianwen Su   University of California at Santa Barbara, USA
Hagen Völzer   IBM Research - Zurich, Switzerland
Matthias Weidlich   Humboldt-Universität zu Berlin, Germany

Program Committee

Ahmed Awad   Cairo University, Egypt
Giuseppe De Giacomo   Sapienza University of Rome, Italy
Jörg Desel   Fernuniversität in Hagen, Germany
Claudio di Ciccio   Vienna University of Economics and Business, Austria
Chiara Di Francescomarino   Fondazione Bruno Kessler-IRST, Italy
Rik Eshuis   Eindhoven University of Technology, The Netherlands
Hans-Georg Fill   University of Bamberg, Germany
Guido Governatori   Data61|CSIRO, Australia
Gianluigi Greco   University of Calabria, Italy
Richard Hull   IBM, USA
Irina Lomazova   National Research University Higher School of
                                     Economics, Russian Federation

Alessio Lomuscio	Imperial College London, UK
Fabrizio Maria Maggi	University of Tartu, Estonia
Andrea Marrella	Sapienza University of Rome, Italy
Heinrich C. Mayr	Alpen-Adria-Universität Klagenfurt, Austria
Oscar Pastor	Universitat Politècnica de València, Spain
Geert Poels	Ghent University, Belgium
Artem Polyvyanyy	Queensland University of Technology, Australia
Wolfgang Reisig	Humboldt Universität zu Berlin, Germany
Arik Senderovich	University of Toronto, Canada
Andreas Solti	Vienna University of Economics and Business, Austria
Ernest Teniente	Universitat Politècnica de Catalunya, Spain
Daniele Theseider Dupré	Università del Piemonte Orientale, Italy
Victor Vianu	University of California San Diego, USA
Lijie Wen	Tsinghua University, China

Track II (Engineering)

Senior Program Committee

Jan Mendling	Vienna University of Economics and Business, Austria
Cesare Pautasso	University of Lugano, Switzerland
Hajo A. Reijers	Vrije Universiteit Amsterdam, The Netherlands
Stefanie Rinderle-Ma	University of Vienna, Austria
Pnina Soffer	University of Haifa, Israel
Wil van der Aalst	RWTH Aachen University, Germany
Boudewijn van Dongen	Eindhoven University of Technology, The Netherlands
Jianmin Wang	Tsinghua University, China
Barbara Weber	Technical University of Denmark, Denmark

Program Committee

Marco Aiello	University of Stuttgart, Germany
Amin Beheshti	Macquarie University, Australia
Andrea Burattin	Technical University of Denmark, Denmark
Cristina Cabanillas	Vienna University of Economics and Business, Austria
Josep Carmona	Universitat Politècnica de Catalunya, Spain
Fabio Casati	University of Trento, Italy
Jan Claes	Ghent University, Belgium
Francisco Curbera	IBM, USA
Massimiliano de Leoni	Eindhoven University of Technology, The Netherlands
Jochen De Weerdt	Katholieke Universiteit Leuven, Belgium
Remco Dijkman	Eindhoven University of Technology, The Netherlands
Marlon Dumas	University of Tartu, Estonia
Schahram Dustdar	Vienna University of Technology, Austria
Gregor Engels	University of Paderborn, Germany
Joerg Evermann	Memorial University of Newfoundland, Canada
Walid Gaaloul	Télécom SudParis, France

Avigdor Gal   Technion, Israel
Luciano García-Bañuelos   University of Tartu, Estonia
Chiara Ghidini Fondazione   Fondazione Bruno Kessler-IRST, Italy
Daniela Grigori   University of Paris-Dauphine, France
Dimka Karastoyanova   University of Groningen, The Netherlands
Christopher Klinkmüller   Data61|CSIRO, Australia
Agnes Koschmider   Karlsruhe Institute of Technology, Germany
Jochen Kuester   Bielefeld University of Applied Sciences, Germany
Henrik Leopold   Vrije Universiteit Amsterdam, The Netherlands
Raimundas Matulevicius   University of Tartu, Estonia
Massimo Mecella   Sapienza University of Rome, Italy
Hamid Motahari   IBM, USA
Jorge Munoz-Gama   Pontificia Universidad Católica de Chile, Chile
Hye-Young Paik   The University of New South Wales, UK
Luise Pufahl   University of Potsdam, Germany
Manuel Resinas   University of Seville, Spain
Shazia Sadiq   The University of Queensland, Australia
Minseok Song   Pohang University of Science and Technology,
                                 South Korea
Stefan Tai   Technical University of Berlin, Germany
Samir Tata   IBM, USA
Arthur Ter Hofstede   Queensland University of Technology, Australia
Farouk Toumani   Blaise Pascal University, France
Moe Wynn   Queensland University of Technology, Australia

Track III (Management)

Senior Program Committee

Joerg Becker   European Research Center for Information Systems,
                                 Germany
Alan Brown   University of Surrey, UK
Mikael Lind   University of Borås, Sweden
Peter Loos   Saarland University, Germany
Amy Looy   Ghent University, Belgium
Olivera Marjanovic   University of Technology, Sydney, Australia
Jan Recker   University of Cologne, Germany
Maximilian Roeglinger   University of Bayreuth, Germany
Michael Rosemann   Queensland University of Technology, Australia
Schmiedel Theresa   University of Liechtenstein, Liechtenstein
Peter Trkman   University of Ljubljana, Slovenia

Program Committee

Peyman Badakhshan   University of Liechtenstein, Liechtenstein
Alessio Braccini   University of Tuscia, Italy
Patrick Delfmann   University of Koblenz-Landau, Germany

Peter Fettke	German Research Center for Artificial Intelligence (DFKI) and Saarland University, Germany
Kathrin Figl	Vienna University of Economics and Business, Austria
Thomas Grisold	Vienna University of Economics and Business, Austria
Marta Indulska	The University of Queensland, Australia
Mieke Jans	Hasselt University, Belgium
Janina Kettenbohrer	University of Bamberg, Germany
John Krogstie	Norwegian University of Science and Technology, Norway
Xin Li	City University of Hong Kong, Hong Kong, SAR China
Alexander Maedche	Karlsruhe Institute of Technology, Germany
Willem Mertens	Queensland University of Technology, Australia
Charles Moeller	Aalborg University, Denmark
Oliver Mueller	IT University of Copenhagen, Denmark
Markus Nuettgens	University of Hamburg, Germany
Ferdinando Pennarola	Università L. Bocconi, Italy
Flavia Santoro	Universidade Federal do Estado do Rio de Janeiro, Brazil
Anna Sidorova	University of North Texas, USA
Silvia Inês Dallavalle de Pádua	University of São Paulo, Brazil
Vijayan Sugumaran	Oakland University, USA
Oliver Thomas	University of Osnabrück, Germany
Harry Wang	University of Delaware, USA
Charlotte Wehking	University of Liechtenstein, Liechtenstein
Axel Winkelmann	University of Würzburg, Germany
Dongming Xu	The University of Queensland, Australia
Weithoo Yue	City University of Hong Kong, Hong Kong, SAR China
Sarah Zelt	University of Liechtenstein, Liechtenstein
Michael Zur Muehlen	Stevens Institute of Technology, USA

Additional Reviewers

Alessio Cecconi	Benjamin Spottke	Fabio Patrizi
Alexey A. Mitsyuk	Bian Yiyang	Fabrizio Maria Maggi
Alin Deutsch	Boris Otto	Florian Bär
Anna Kalenkova	Brian Setz	Francesco Leotta
Anton Yeshchenko	Carl Corea	Frank Blaauw
Armin Stein	Chiara Ghidini	Friedrich Holotiuk
Azkario Rizky Pratama	Christoph Drodt	Giorgio Leonardi
Bastian Wurm	Daning Hu	Gottfried Vossen
Benjamin Meis	David Sanchez-Charles	Gul Tokdemir

Harris Wu
Heerko Groefsema
Jennifer Hehn
Jiaqi Yan
Kieran Conboy
Kimon Batoulis
Kun Chen
Laura Giordano
Lele Kang
Luciano García-Bañuelos
Mauro Dragoni

Michael Leyer
Montserrat Estañol
Nils Urbach
Peyman Badakhshan
Rene Abraham
Riccardo De Masellis
Robin Bergenthum
Roope Jaakonmäki
Saimir Bala
Sebastian Steinau
Sergey Shershakov

Shan Jiang
Stefan Oppl
Sven Radszuwill
Thomas Friedrich
Tyge-F. Kummer
Vladimir Bashkin
Vladimir Zakharov
Wei Wang
Xavier Oriol
Yuliang Li

Sponsors

Conference Partner

Platinum Sponsor

Gold Sponsor

Gold Sponsor

Bronze Sponsor

Contents

Track III: Management

Track I: Foundations

Process Mining Crimes – A Threat to the Validity of Process Discovery Evaluations

Jana-Rebecca Rehse$^{(\boxtimes)}$ and Peter Fettke

Institute for Information Systems (IWi) at the German Center for Artificial
Intelligence (DFKI GmbH) and Saarland University,
Campus D3 2, Saarbrücken, Germany
{Jana-Rebecca.Rehse,Peter.Fettke}@iwi.dfki.de

Abstract. Given the multitude of new approaches and techniques for
process mining, a thorough evaluation of new contributions has become
an indispensable part of every publication. In this paper, we present a
set of 20 scientifically supported "process mining crimes", unintentional
mistakes that threaten the validity of process discovery evaluations. To
determine their prevalence even in high-quality publications, we perform
a meta-evaluation of 21 process discovery papers published at the BPM
conference. We find that none of these papers is completely crime-free,
but the number of crimes and their impact on the evaluations' validity
differs considerably. Based on our list of crimes, we suggest a catalog of
13 process mining guidelines, which may contribute to avoiding process
mining crimes in future evaluations. Our objective is to spark an open
discussion about the necessity of valid evaluation results among both
process mining researchers and practitioners.

Keywords: Process mining · Process discovery · Evaluation
Quality metrics

1 Introduction

Process mining is set out to gain insights into information systems by analyzing
their behavior, as recorded in event logs. More specifically, the goal of process
discovery is to represent the behavior of the information systems in form of a
business process model [1, p. 163ff.]. The quality of process discovery results
is often measured in terms of the four dimension fitness (the model's ability to
replay observed behavior), precision (the model's ability to not allow unobserved
behavior), generalization (the model's ability to explain unobserved behavior),
and simplicity (the model's complexity) [2]. Over the last fifteen to twenty years,
a number of process discovery approaches have been proposed [3]. They either
address so far unresolved challenges, such as duplicate tasks, or improve the
state-of-the-art in terms of result quality as measured by the four dimensions or
efficiency, i.e. using less computational resources [4].

© Springer Nature Switzerland AG 2018
M. Weske et al. (Eds.): BPM Forum 2018, LNBIP 329, pp. 3–19, 2018.
https://doi.org/10.1007/978-3-319-98651-7_1

Independent from the goal of an individual approach, an empirical evaluation of new contributions is nowadays an indispensable prerequisite for publication, since it enables reviewers and readers to empirically verify that the claimed contributions have actually been achieved. An evaluation must be complete, relevant, sound, and reproducible, such that it produces scientifically substantiated results. Hence, the validity of the evaluation and therewith the reliability and credibility of the published research results are potentially threatened by erroneously conducted evaluations.

In this paper, we introduce the notion of "process mining crimes", eminent but unintentional mistakes that impact the validity of an evaluation in the process mining domain. Although we use the hyperbolic expression "crime", we do not insinuate any criminal intentions, but follow the original terminology by Heiser [5] and van der Kouwe et al. [6]. We give a first list of six crime categories and 20 individual crimes and assess their prevalence in 21 previous BPM conference publications on process discovery. In the process, we develop a better understanding of which evaluation aspects are generally well accounted for and which aspects need additional attention. This paper is meant to sharpen the process mining community's awareness for the importance of empirical evaluations and the underlying assumptions and scientific principles. Ideally, it will spark a constructive and critical discussion, which may lead towards explicit guidelines for process mining evaluations. It is explicitly not our intention to point fingers or blame individual researchers for past mistakes. The problem we stress here does not concern individual papers as much as the field as a whole, which is why we do not point out crimes in individual papers. We also refrain from speculating about potential reasons for committing these crimes and believe that most, if not all, identified crimes were unintentional instead of deliberate.

Our paper has the following structure. We introduce 20 process mining crimes in Sect. 2 and examine their prevalence in a meta-evaluation of 21 BPM conference papers on process discovery in Sect. 3. To prevent future process mining crimes, we propose 13 process mining evaluation guidelines in Sect. 4. We report on related work in Sect. 5. Section 6 concludes the paper with a discussion on limitations and implications of our research.

2   Process Mining Crimes

Evaluations pursue three major objectives. For the practitioner, they demonstrate a technique's ability to solve a given problem, indicating whether to adopt it into organizational practice. For the researcher, they attest a new technique's superiority over the current state-of-the-art, either by solving a known problem better or more efficiently or by presenting the first solution to a previously unresolved problem. For the research community, they empirically verify a scientific contribution, adding to the body of knowledge in the given field. In order to fulfill these objectives, evaluations have to meet four requirements [6].

- *Completeness*, i.e. they must separately address each claimed contribution and also report on any negative impact.
- *Relevance*, i.e. they must provide meaningful and truthful information.

- *Soundness*, i.e. the employed measurements must be accurate, valid, and reliable regarding the indicators to be measured.
- *Reproducibility*, i.e. sufficient information must be provided for others to repeat the evaluation and roughly achieve the same results.

A good evaluation in any empirical research discipline should fulfill these criteria to ensure its scientific validity. Process discovery evaluations are no exception, as they provide conclusive evidence about a new discovery techniques. An incomplete evaluation may leave out measurements such as model quality or runtime, making it difficult to compare approaches. An irrelevant evaluation may contain misleading measurements that are not applicable to asses an approach's capabilities. Unsound evaluation results may have been incorrectly measured or computed, misrepresenting an approach's capabilities. Irreproducible evaluation results cannot be replicated, contradicting the scientific method and impeding practical applicability, e.g. in case of missing hardware resources.

In practice, however, many existing publications commit "process mining crimes", which violate the requirements. In this section, we present a list of 20 process mining crimes, which we deem important enough to diminish the validity of an evaluation in a paper on process discovery. It is based on two contributions on "benchmarking crimes" in system security [6] and operation systems [5], but we adapted and regrouped the crimes to be relevant for process discovery. Table 1 lists 20 crimes in six categories, and each crime's impact on the above requirements. For better reference, categories are numbered and crimes are addressed by a code (category number + letter).

Category 1 addresses the choice of evaluation data, which influences evaluation completeness, relevance, and soundness. The characteristics of an event log are a main factor in assessing the quality of a discovery techniques [7]. Choosing the wrong evaluation data may render an evaluation void, if the data does not exhibit the technique's required characteristics. Hence, researchers must pay close attention to their data selection and may not (1a) *choose their evaluation data without proper justification*. Also, they may not (1b) *use micrologs for representing overall performance results*. Micrologs are small logs used for measuring the ability of a discovery technique to handle specific aspects, e.g. loops or duplicate tasks. While they may be helpful in many situations, an evaluation cannot be based on micrologs alone, but instead must assess a technique under more realistic circumstances. The same holds true for (1c) *evaluating simplified simulated logs*, which were generated solely for the purpose of evaluation. As their characteristics are generally not representative for process logs, their results may not be generalized. In general, the evaluated log must be designed in a way, such that it is suitable to measure a technique's improvement over the state-of-the-art. (1d) *Inappropriate or misleading logs* may not be used.

Category 2 deals with assessing the quality of a discovered model according to the four dimensions fitness, precision, generalization, and simplicity. They represent the counteracting objectives of process discovery. As these dimensions are rather informal, they can be measured by multiple metrics. Slight differences between the metrics result in different quality assessments, over- or underestimating the perceived quality of a model [8]. To account for these deviations,

Table 1. Process mining crimes and their impact on completeness (Compl.), relevance (Rel.), soundness (Sound), and reproducibility (Repr.)

Process mining crimes		Compl.	Rel.	Sound	Repr.
1	*Using the wrong evaluation data*				
1a	Choice of evaluation data without proper justification	•	•		
1b	Micrologs representing overall performance		•	•	
1c	Evaluation of simplified simulated logs		•	•	
1d	Inappropriate and misleading logs		•		
2	*Misleading quality assessment*				
2a	Selective quality metrics hiding deficiencies	•			
2b	Matching quality metrics		•		
2c	Only measure selective quality dimensions	•			
3	*Scientific inaccuracies*				
3a	Not evaluating potential quality degradation	•			
3b	Creative evaluation result accounting			•	
3c	Not all claims empirically verified	•			
4	*Incomplete evaluations*				
4a	No indication of significance of results	•		•	
4b	No assumptions on noise	•		•	
4c	Elements of approach not tested incrementally	•			
5	*Improper comparison of evaluation results*				
5a	No proper comparison		•		
5b	Only evaluate against yourself		•		
5c	Unfair evaluation of competitors		•		
6	*Missing information*				
6a	Missing hardware specification				•
6b	Missing software specification				•
6c	Individual measures not given	•			•
6d	Relative numbers only	•			•

evaluations may not use (2a) *selective quality metrics hiding deficiencies*, i.e. use metrics that are unsuitable for measuring the model quality. Ideally, authors either justify the selected metrics' measurement precision or evaluate a quality dimension with several metrics. Similarly, authors may also not use (2b) *matching quality metrics*, i.e. closely related quality metrics that e.g. use the same technical concept to measure different quality dimensions, as they can hide a techniques's deficiencies [9]. Different quality metrics should therefore be selected from multiple independent sources. Finally, evaluations may not (2c) *measure selective quality dimensions*. There is no single number that is able to assess the overall performance of a process discovery technique, which is why the four dimensions have been established. Process discovery quality assessments always affect several dimensions, but also factors like runtime [3]. Hence, it is unacceptable to only measure selected quality dimensions without proper justification.

Category 3 covers rather generic crimes adapted to process discovery. The first crime addresses the full disclosure of all empirical insights into a discovery technique and their implications to its practical application. As we explain above, process discovery quality is typically assessed in terms of four dimensions, because there is no single number that is able to assess the overall performance of a process discovery technique. Increasing one quality dimension may cause another to decrease, since most discovery techniques are unable to address all dimensions at once [2]. This is generally acceptable, as technique may be set out to prioritize one quality dimension (e.g. precision) over the others. However, by (3a) *not evaluating potential quality degradation*, researchers fail to disclose the implications such a prioritization may have on the other quality dimensions (e.g. generalization). Other researchers cannot identify further research gaps to close and practitioners may apply discovery techniques that are not fitting their process mining objective, hence not providing any business value.

Crime (3b) *creative evaluation result accounting* has similar implications, but they are generally harder to assess, as there are many different ways to inaccurately report on or explain evaluation results. Examples include incorrect computations or a lack of result explanation, but in general, all statements that cannot be directly inferred from the given evaluation results. Finally, papers may only claim results which they are empirically proving. If (3c) *not all claims are empirically verified*, readers can be mislead into attributing a technique with quality results it did not achieve. This hinders the progress of the field, as future papers addressing these claims may be prevented from publication.

Category 4 contains crimes related to evaluation completeness. Evaluations may be correct in all other aspects, but if critical steps or results are missing, they cannot be completely trusted. This may concern performed measurements, particularly if there is (4a) *no indication of result significance*. While many quality metrics are assessed by algorithms, other measurements (such as runtimes or heuristics) are subject to random fluctuations. To realistically assess the explanatory power of such computations, it is not enough to perform only one measurement, as statistical outliers may occur. Instead, researchers must perform multiple runs and report on the result significance, e.g. give the standard deviation, such that readers know how reliable the given numbers actually are.

Similarly, an evaluation may not contain (4b) *no assumption on noise*, since the amount of noise in an event log impacts the significance of a quality assessment [4]. Without knowing the assumed or expected amount of noise, it is impossible to draw conclusions about the reliability of a quality assessment, as noise may lead to deviations of five to ten percentage points in many popular quality metrics. Finally, process discovery techniques may contain several at least partially independent elements, which may each have a decisive influence on the result. If (4c) *elements of an approach are not tested incrementally*, their individual impact may not be measured adequately, such that quality improvements can be attributed correctly. It remains unclear, which elements of a newly presented technique are actually necessary to achieve a better result quality.

Category 5 focuses on comparisons between process discovery techniques. A paper's contribution can often only be assessed when comparing it to the existing state-of-the-art to find out whether the new approach is in fact able to produce better or faster results. If those comparisons are not properly executed, the results cannot be trusted. Therefore, a discovery technique that claims to be an improvement over existing solutions must always contain an appropriate comparison with a state-of-the-art competitor. If there is (5a) *no proper comparison*, it is impossible to realistically assess its contribution. Also, if papers claim to address a previously unresolved problem, they must clearly explain where they surpass the state-of-the-art and why a comparison would not make sense.

When selecting techniques to compare with, it is unacceptable to (5b) *only evaluate against yourself.* If researchers compare new solutions to their own earlier work instead of state-of-the-art competitors, the results will lack relevance. Finally, in a comparison, there is always a risk of (5c) *unfair evaluation of competitors*, i.e. applying a competing technique such that it does not produce optimal results. Competitors can e.g. not be given the requisite amount of computing resources, or be configured suboptimally. Results from such a comparison can be seen as irrelevant at best and deceitful at worst.

Category 6 refers to missing information, i.e. crimes that fail to fully disclose all relevant information about the evaluation. This threatens both completeness and reproducibility. The latter is particularly important, as intersubjective confirmation of research results is one of the main properties of scientific inquiry. Therefore, (6a) *missing hardware specification* invalidates all statements regarding a technique's performance or practical applicability, as these depend on the computer on which a discovery has been executed. Similarly, (6b) *missing software specifications* impairs reproducibility of both technique and evaluation. Readers of process discovery papers must be provided enough information to perform a replication study, i.e. reproduce the evaluation results by either reimplementing an approach or using a provided software. This requires authors to fully disclose both the details of their contribution and the specifics of the evaluation. Self-developed software should ideally be made publicly available. For other software (e.g. frameworks like ProM, but also dependent packages), the version numbers must be given. An evaluation can only be fully replicated when all results are known to the reader. If (6c) *individual measures are not given* when evaluating multiple logs and/or quality metrics, not only can the differences between the separate measurements not be assessed to allow for conclusions about the strong and weak points of the technique, it is also impossible to confirm the correctness of a replication. The same holds true for comparing discovery results to those of other techniques. If (6d) *only relative numbers* are given instead of absolute ones (e.g. percentage of correctly discovered models), it is impossible to assess the results' plausibility.

Both categories and crimes are based on general considerations on the scientific process as well as empirical studies on the quality assessment of process discovery techniques, so their frequencies remain unclear. We generally believe that each committed crime is a crime too much, but future research activities should focus on the more frequent and therefore higher-impact crimes. So, in the following section, we evaluate process discovery papers from earlier BPM conferences regarding the prevalence of committed crimes.

3 Meta-evaluation

3.1 Objectives and Paper Selection

In order to gain a better understanding of the process mining crimes' practical relevance, we conduct a meta-evaluation to determine their prevalence in high-quality papers on process discovery. We do not intend to analyze every process discovery paper ever published, instead, our goal is to raise the community's awareness regarding the impact of process mining crimes to our everyday research. Therefore, we limit our scope to the BPM conference, which due to its role as one of the major publication outlets for process mining and its very high quality standards should provide an compelling sample to assess the crimes' severity and frequency. We inspected the proceedings of every BPM conference (2003–2017) and assessed the 418 published papers according to the following criteria. This resulted in a total of 21 papers, presented in Table 2.

- The paper presents a technique that transforms an event log (input) into a procedural process model (output).
- The paper makes a contribution to process discovery, i.e. generating a process model that represents the input log behavior in the most appropriate way.
- The paper contains an empirical evaluation.

3.2 Codification Process

To ensure the validity and reproducibility of our research, we developed codification guidelines, i.e. the "evidence" required to determine whether or not a paper commits a crime. These guidelines were adjusted after gaining some practical experience with their application and an in-depth discussion between the two researchers, who independently evaluated the papers. Some crimes required "positive evidence", i.e. information that had to be present for the crime to be committed. For example, in order to codify crime 2a (selective quality metrics), papers needed to specify the quality metrics they used in their evaluation. Other crimes could only be assessed by "negative evidence", i.e. the absence of information. For example, crime 4b (no assumptions on noise) was codified as committed, if no explicit or implicit (e.g. filtering of infrequent behavior) assumption on noise could be found in the evaluation section. For each paper, each crime was codified in four stages.

- *Applies:* Based on the provided information, the paper commits this crime.
- *Flawed:* Whether or not the paper commits this crime cannot be finally decided due to imprecise or missing information.
- *Correct:* The paper does not commit this crime.
- *Not applicable:* The paper could not have committed this crime.

Table 2. Evaluated papers

2003	Golani et al.: Generating a Process Model from a Process Audit Log
	Schimm: Mining Most Specific Workflow Models from Event-Based Data
2007	Ferreira et al.: Approaching Process Mining with Sequence Clustering: [...]
	Günther et al.: Fuzzy Mining - Adaptive Process Simplification Based on [...]
2008	Carmona et al.: A Region-Based Algorithm for Discovering Petri Nets [...]
	Diniz et al.: Automatic Extraction of Process Control Flow from I/O [...]
2009	Carmona et al.: Divide-and-Conquer Strategies for Process Mining
	Bose et al.: Abstractions in Process Mining: A Taxonomy of Patterns
2013	Buijs et al.: Mining Configurable Process Models from Collections [...]
	Ekanayake et al.: Slice, Mine and Dice: Complexity-Aware Automated [...]
2014	Conforti et al.: Beyond Tasks and Gateways: Discovering BPMN Models [...]
	Vazquez-Barreiros et al.: A Genetic Algorithm for Process Discovery [...]
	Redlich et al.: Constructs CompetitionMiner: Process Control-Flow [...]
2015	Guo et al.: Mining Invisible Tasks in Non-free-choice Constructs
	Ponce-de-Léon et al.: Incorporating Negative Information in Process Discovery
	van Zelst et al.: Avoiding Over-Fitting in ILP-Based Process Discovery
	Liesaputra et al.: Efficient Process Model Discovery Using Maximal [...]
2016	Lu et al.: Handling Duplicated Tasks in Process Discovery by Refining [...]
	de San Pedro et al.: Discovering Duplicate Tasks in Transition Systems [...]
2017	van der Aalst et al.: Learning Hybrid Process Models from Events [...]
	Chapela-Campa et al.: Discovering Infrequent Behavioral Patterns in [...]

For example, a paper was codified as committing crime 2a if a quality dimension is measured by one metric only or several metrics that are based on the same underlying technique. If the process quality is not numerically assessed, it was codified as not applicable. If the assessed dimensions are measured by at least two different metrics (e.g. by averaging the results), it was codified as correct. If the given information is not sufficient to determine which metrics and dimensions are assessed, it was codified as flawed. Crime 4c was not applicable if the approach does not consist of multiple parts. Crimes 5b and 5c were not applicable if the paper did not contain a comparison with another approach.

For our codification, we used only the information provided in the paper itself and, wherever possible, directly related additional material (e.g. cited technical reports including all empirical data), essentially taking on the perspective of a reviewer in a peer-review process. As we have stated above, we decided not to include the codification of the individual papers here, as to avoid blaming. However, we are more than open to provide both the codification guidelines and the individual paper assessments to interested readers upon request.

3.3   Results

Table 3 summarizes the results of our evaluation. For each crime, we report on the number of papers where said crime applies (A), that are flawed (F), and this crime is not applicable (N). From that, we compute the crime ratio (CR), i.e. the percentage of papers where this crime is applicable and either committed or flawed ($\frac{A+F}{21-N}$) and the applicability ratio (AR), i.e. the percentage of papers where this crime applies ($\frac{21-N}{21}$). The crime ratio determines the severity of a crime. The higher this number, the more papers are not flawless regarding the requirements stated in the crime description. To get more accurate results, the number of crime-committing papers is set in relation to the number of papers where this crime is applicable, i.e. those papers where it would have been impossible to commit said crime, are not considered in the CR. This is why the applicability ratio is helpful, as it determines a crime's potential frequency, i.e. the ratio of papers where this crime could have been committed. As such, the AR provides more context for the crime ratio.

Table 3. Evaluation results with number of papers where the crime applies (A), is flawed (F), is not applicable (N), crime ratio (CR), and applicability ratio (AR)

Process mining crimes		A	F	N	CR	AR
1a	Choice of evaluation data without proper justification	2	5	0	33%	100%
1b	Micrologs representing overall performance	3	2	0	24%	100%
1c	Evaluation of simplified simulated logs	4	2	0	29%	100%
1d	Inappropriate and misleading logs	0	1	0	5%	100%
2a	Selective quality metrics hiding deficiencies	8	1	6	60%	71%
2b	Matching quality metrics	3	3	9	50%	57%
2c	Only measure selective quality dimensions	6	0	7	43%	67%
3a	Not evaluating potential quality degradation	4	1	1	25%	95%
3b	Creative evaluation result accounting	0	1	1	5%	95%
3c	Not all claims empirically verified	2	3	0	24%	100%
4a	No indication of significance of results	11	0	7	79%	67%
4b	No assumptions on noise	7	8	1	75%	95%
4c	Elements of approach not tested incrementally	0	0	15	0%	29%
5a	No proper comparison	0	0	9	0%	57%
5b	Only evaluate against yourself	0	0	10	0%	52%
5c	Unfair evaluation of competitors	0	4	10	36%	52%
6a	Missing hardware specification	15	3	0	86%	100%
6b	Missing software specification	4	9	0	62%	100%
6c	Individual measures not given	2	0	4	12%	81%
6d	Relative numbers only	2	1	4	18%	81%

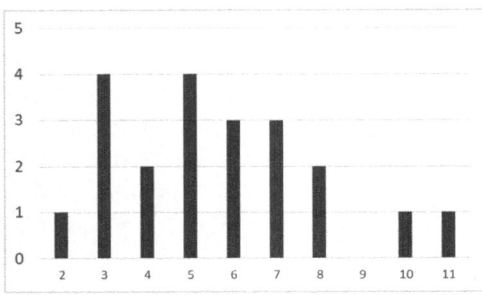

Fig. 1. Papers per number of crimes

Figure 1 shows a histogram on the number of crimes per paper. The average number of crimes per paper is about 5.6, with a minimum of 2, a maximum of 11, and a standard deviation of 2.3. As we can see from the histogram, the number of crimes per paper is distributed unevenly. We can already state that there is no paper that is completely free of crimes, and there are only very few papers that commit an exceptionally high number of crimes.

4   Process Mining Guidelines

In this section, we propose 13 process mining guidelines in order to avoid process mining crimes in future evaluations. Therefore, we examine each crime and derive a course of action that we believe will improve the quality of process discovery evaluations. Since guidelines are only useful if they are the generally-agreed-upon result of a community-wide discussion, we see this a first suggestion that is to be further discussed and refined.

The choice of evaluation data is a major factor in evaluating a process discovery technique, hence the high number of related crimes (1a, 1b, 1c, 1d). Selecting the wrong input log for a process discovery evaluation (e.g. applying a technique to detect duplicate tasks to a log without duplicate tasks) invalidates all evaluation results, independent from all other choices. Crime 1a concerns the justification of the evaluation data selection, which affects at least a third of our evaluated papers. Often, researchers evaluate "real-life logs" without any additional justification relating to their concrete technique. In our opinion, this is not sufficient, as logs generated by practical systems may not contain "real-life features" (such as noise, loops, high data volumes, duplicate tasks, etc.). Authors should at least specify and check in which way their chosen data is representative to demonstrate their technique's capabilities. Another possible justification is that data such as the BPI challenge logs are frequently used by other papers, allowing for a better comparison to the state-of-the-art.

Crimes 1b, 1c, 1d directly address the unrepresentative nature of certain logs. It is mainly due to misleadingly generalizing evaluation results gained from micrologs (small, artificial logs that contain one predominant feature) or logs that were simulated specifically for a certain evaluation. Both log types can be useful, especially when verifying a technique's capability to handle the contained features (such as duplicate tasks) to avoid crime 1d. However, the results on those logs may not allow assessing the technique in general, because they do not contain other behavior that can be assumed to interfere with the result quality. So, an evaluation with an artificially generated log should ideally be accompanied by another evaluation with a second external log containing e.g. the real-life features mentioned above, such that results can be generalized. As a conclusion, crimes from category 1 can be avoided by following **Guideline 1: Choose representative evaluation data and justify this choice**.

Crimes from category 2 rely on the four quality dimensions (fitness, precision, generalization, and simplicity) and the numerous metrics that exist for assessing them. As [8,9] show, these metrics tend to over- und underestimate the perceived values of these dimensions, since certain algorithmic approaches are incapable of factoring in certain log characteristics. This might also happen, if two different dimensions are assessed with metrics based on the same technique. Therefore, evaluation validity can be improved by **Guideline 2: Measure each quality dimension with multiple metrics**. Ideally, they are based on different techniques by different authors.

The objective of a discovery technique cannot always be properly assessed with all four dimensions, as there are approaches that deliberately focus on improving a subset of them. This does not mean that the other dimensions may be simply omitted as crime 2c states; a model that focuses on only one or two dimensions is often not useful. However, in some cases, evaluating all four dimensions does not provide any additional value to the evaluation, because e.g. simplicity does not directly relate the model to the log [4]. In general, researchers should follow **Guideline 3: When omitting quality dimensions from the evaluation, this choice must be properly justified**.

No researcher likes to admit flaws, drawbacks, or unresolved issues in newly presented research. There is also the common impression among researchers that pointing towards the negative sides of one's research will diminish the chances to be accepted at high-quality conferences. While we cannot reject this as untrue, we want to encourage to handle this sort of "publication bias" more openly. Researchers should report both the assets and drawbacks of their contribution, such that the scientific community is able to assess which new knowledge has been developed and which problems remains unsolved, motivating more research. Likewise, reviewers should acknowledge honesty and judge a paper on its novelty and contribution instead of the vast number of problems it does not solve.

The necessity for this kind of scientific demeanor is motivated by the high applicability ratio, high impact, and notable crime ratio of the concerned crimes 3a, 3b, and 3c. Crime 3a concerns balancing the different quality dimensions in process discovery. As [2] states, discovery techniques are typically unable to address all four quality dimensions plus non-functional quality aspects such as runtime or memory consumption at once, given the contradictory nature of e.g. precision and generalization. So, a new approach may prioritize reaching high values in a subset of quality dimensions and therefore accept lower values in others. This is a valid research objective, however, openly communicating these choices is necessary to correctly assess the abilities of an approach and compare them to the goals of a concrete process discovery project.

Crimes 3b and 3c have a high impact, as they potentially associate a paper with a contribution it does not make, impeding the publication of future research on the same topic. They are not particular to process discovery, but apply to all empirical sciences, making them all the more relevant. However, accurately accounting for evaluation results and carefully describing the contributions of a paper without uncalled-for generalization is easily achievable, reducing the risks associated with these crimes. Together with crime 3a, this motivates **Guideline 4: Be honest about a technique's contributions and weak points, to allow future research**.

Crime 4a does not affect all evaluations, only those that perform "real" measurements, i.e. assign a number to an evaluated property by means of a unit (e.g. runtime measured in milliseconds) and those that use non-deterministic computations. Reporting on the variance of these numbers is crucial to account for statistical outliers, random choices, or the imprecision of the measuring instruments. Hence, **Guideline 5: When reporting on measurements or non-deterministic computations, perform several runs and report on the statistical significance**.

Empirical research has shown that noise in the evaluated log may reduce the precision of the evaluation results, such that they have to be taken with a grain of salt [4]. Therefore, to put a perspective on evaluation results, there must be a discussion about the assumed noise. Explicitly defining a noise level in a "real-world log" created by a productive system might be difficult, but at least discussing it is part of explicating the evaluation log features, as addressed by crime 1a above. Another option would be to remove aborted instances or other atypical behavior or to justify why the discovered model benefits from including it. Concluding, to avoid committing crime 4b, researchers must follow **Guideline 6: Explicate assumptions about noise in the evaluation data**.

Crime 4c addresses techniques that combine different parts to achieve an improvement over the state-of-the-art. They need to be evaluated separately, such that their impact on the process discovery quality can be discerned. As a rule of thumb, if a technique consists of multiple parts that can be reasonably separated from one another, such that applying them still yields a process model, follow **Guideline 7: Include a separate evaluation for each step**.

As crime 5a states, the contribution of a new technique is only identifiable when comparing it to the existing state-of-the-art. This is particularly true for techniques that claim to solve a problem in a better or more efficient way, e.g. discovering process models with better precision values or less memory consumption, but it is also relevant for techniques that claim to address a previously unresolved problem. While, in these cases, it might not make sense to compare the numerical quality of discovered models, it is indispensable to identify existing work and explain how the new contribution differs and goes beyond existing approaches to related problems, hence **Guideline 8: Include a comparison**.

The expressiveness of a comparison is determined by the compared techniques. Often, researchers work on a technique over a long period of time and develop it to address its weaknesses. So, when publishing a new and improved version of an existing technique, it is enticing to compare it to its predecessor such that the improvements can be made clear. But one's own previous work may not be state-of-the-art anymore, because other researchers have published better techniques in the meantime and such a comparison is nothing more than a self-reference. Hence, comparing an approach should always consider **Guideline 9: Compare to state-of-the-art techniques by other researchers**.

When working with other researcher's techniques, it is the easiest to use an existing open-source implementation. If such an implementation does not exist (sometimes authors make it available upon request), it needs to be re-implemented exactly as specified in the given contribution. Crime 6b is concerned with presenting a technique in a re-implementable way, especially pointing out the necessary parameters. Our most commonly encountered evidence for crime 5c (unfair evaluation of competitors) concerned suboptimal configurations of competing approaches. We acknowledge that especially when applying several competing techniques to multiple logs, it is not always feasible to find an optimal configuration for each of them. However, it is not a valid evaluation, if one's own approach is used with optimal configurations, while others are not. So, in order to guarantee a fair comparison, the same configuration settings (optimized or default) should be applied to all approaches, such that **Guideline 10: Comparisons are executed fairly and appropriately**.

Finally, an evaluation must always be as honest and as complete as possible to enable other researchers to thoroughly understand and replicate its findings. Therefore, to counteract crimes 6a and 6b, follow **Guideline 11: Specify the hardware used for experiments** and **Guideline 12: Provide the implementation source code used for the experiments**. If this is impossible, make sure to describe all parameters necessary to re-implement an approach. Crimes 6c and 6d are avoided by **Guideline 13: Provide full access to all evaluation data**. If the data is too large to be included in the paper itself, publish it online and provide a link.

5   Related Work

Over the last almost 20 years, a number of process discovery techniques have been proposed [3]. After essential challenges such as handling duplicate tasks or non-free choice constructs have been tackled, newer papers are more focused on outperforming each other in terms of quality or efficiency. This development elicits the need for commonly accepted and scientifically valid evaluation methods [4]. Recently, there have been activities towards the development of a structured evaluation framework for process discovery, which may help avoid process mining crimes. The first suggestion was made by Rozinat et al., who define quality dimensions along with metrics and other framework components [10,11]. Weber et al. present a short and concise procedure model for experimental evaluations and comparisons between process mining techniques [12]. The CoBeFra framework offers an extensible architecture, along with a ProM implementation that can be used for performing empirical computations [13].

Among process mining researchers, there is a certain consensus on the suitability of the four dimensions fitness, precision, generalization, and simplicity for measuring the quality of a discovered process model with respect to the original log [11]. There are empirical works investigating the impact of said consensus on the validity of process quality evaluation. The evaluation by Buijs et al. shows that most process discovery approaches are not able to address all quality dimensions simultaneously and also do not allow to weigh their relevance in a concrete mining scenario [2]. In [7], vanden Broucke et al. examine how event log characteristics influence the performance of discovery techniques.

Each dimension can be assessed by different metrics with different approaches to operationalize the informal notions behind the dimension. Other empirical works study the impact of the chosen metrics on process discovery evaluations. A good overview on current metrics for measuring fitness, precision, and generalization is given by Janssenwillen et al. [9]. From their comparative study, they conclude that fitness and precision metrics are fairly consistent, although there is some variety regarding their sensitivity, and computational feasibility plays a major role. Generalization metrics, however, are not consistent with each other, requiring substantial additional research. In [8], Tax et al. define requirements for precision measures and come to the (contradictory) conclusion that none of the existing metrics is able to measure precision in a consistent and valid way.

Some recent insights on evaluation validity differentiate between model, log, and system [2]. Depaire suggests a statistical measure to determine the likelihood that a discovered model actually represents the unknown underlying system [14]. The study by Janssenwillen et al. shows that current fitness and precision metrics are significantly biased by log incompleteness and noise, limiting their ability to assess the model quality with respect to an unknown system [4]. Our own study on the influence of unobserved behavior on process discovery quality indicates that the unknown system nature and hence the amount of unobserved behavior can have a significant impact on the quality assessment [15].

6 Limitations and Conclusion

In this paper, we introduce process mining crimes, mistakes that threaten the validity of process discovery evaluations. We present a list of 20 crimes and analyze their prevalence in 21 BPM papers on process discovery. Based on this meta-evaluation, we suggest and discuss a catalog of 13 guidelines that are supposed to support scientifically valid process discovery evaluations in the future.

As this is an empirical research paper, we can be held accountable against our own list of crimes. Our meta-evaluation is flawed regarding crimes 1b (micrologs) and 1d (inappropriate logs), given that it contains only a small set of conference papers, which due to scope and limitations do not exhibit the same scientific rigor as journal publications. Because we limited our selection to BPM papers, we did not consider some high-impact process discovery papers, such as the flexible heuristics miner [16], meaning that our selection is not comprehensive. Besides, our paper selection was not externally reproduced, so we potentially missed relevant contributions. We also commit crime 4a, as our set of papers is too small to measure statistical significance.

Although our requirements were carefully selected and are based on empirical research, both the requirements and the crimes are neither sufficient nor carved in stone, hence flawing our paper regarding crimes 2a and 2c. Papers ranked as crime-free might still not have a valid evaluation. Other experts might argue that we have misjudged the impact of crimes on our list, while missing out on other important crimes. Our research is based on empirical work in system security and influenced by our own view as process mining researchers. The impact of a concrete crime in practice could differ from our understanding. Also, due to the differing crime specificity, some crimes may be merged or separated. We actively invite the community of both researchers and practitioners to critically assess our findings and enter a scientific discourse with the overall objective to improve the quality of our research by providing more convincing evaluations.

With the collaboration of two independent researchers, our results are inter-subjectively confirmed, but not yet universally accepted, leaving room for discussion. Due to our decision to not point out specific crimes in concrete papers, we cannot make our research process fully transparent, but we will gladly make our evaluation results available upon request.

An important finding was that some crimes are much easier to identify than others. For example, it is simple to state whether or not a paper has correctly specified the used hardware or made an assumption on noise. Judging whether an evaluation has unfairly evaluated its competitors or chosen an inappropriate evaluation log is much more difficult or even impossible to prove beyond reasonable doubt. This means that we could have potentially missed some crimes, so the crime rate given in Table 3 should rather be seen as a lower bound. A crime that is either very rare or very difficult (or impossible) to prove will have a low crime ratio, however, a low ratio does not mean that this crime is insignificant.

An evaluation that commits a certain crime is also not necessarily of lesser quality than the one where the same crime is inapplicable. For example, a paper focused on producing a maximally precise process model might exclude the simplicity dimension, but fail to justify this decision, hence committing crime 2c. The evaluation might still verify the technique's capability to produce a model with a high precision and therefore confirm its contribution. Meanwhile, the evaluation of a paper where 2c is not applicable does not measure any quality dimensions, limiting the comparability of the evaluated technique against others.

In this context, we also notice that committing crimes is sometimes inevitable for a researcher. Strict space limitations do not always allow for full-fledged evaluations. Some research papers put their focus on one specific quality feature, so the evaluation is positioned accordingly. Micrologs or simulated logs may be necessary if no real-world data is available or suitable to the specifically evaluated features. We acknowledge all of these reasons, but would like to encourage researchers to be explicit about these limitations.

Our list of crimes is fairly closely aligned with the notion of measuring process quality in terms of the four dimensions fitness, precision, generalization, and simplicity [1]. However, these dimensions are not incontrovertible. Particularly, the vague definition of generalization and the ensuing small number of appropriate metrics has drawn criticism [4, 14, 15]. We have still decided to build our list of crimes along this evaluation framework, as it is widely used within the community. We acknowledge that crimes related to quality dimensions or quality metrics (2a, 2b, 2c) may loose significance, if an approach's quality is not measured according to this understanding of process discovery quality.

This paper can be the starting point of research activities that may help to prevent future process mining crimes. As a next step, our crimes should be justified by means of a literature review, to be grounded on empirical findings. Future crime analyses should also compare crime prevalence in conferences versus journal papers, as those typically have more space, but also higher requirements regarding scientific rigor. Also, the individual crimes' impact should be further investigated, to decipher between high-impact and low-impact crimes. Additional research is also required to investigate what causes crimes and how crimes correlate with other publication features. All of this future work will be necessary to better understand the role process mining crimes play in our research.

References

1. van der Aalst, W.: Process Mining: Data Science in Action. Springer, Heidelberg (2016). https://doi.org/10.1007/978-3-662-49851-4
2. Buijs, J., van Dongen, B., van der Aalst, W.: Quality dimensions in process discovery: the importance of fitness, precision, generalization and simplicity. Int. J. Coop. Inf. Syst. **23**(01), 1440001 (2014)
3. De Weerdt, J., De Backer, M., Vanthienen, J., Baesens, B.: A multi-dimensional quality assessment of state-of-the-art process discovery algorithms using real-life event logs. Inf. Syst. **37**(7), 654–676 (2012)
4. Janssenswillen, G., Jouck, T., Creemers, M., Depaire, B.: Measuring the quality of models with respect to the underlying system: an empirical study. In: La Rosa, M., Loos, P., Pastor, O. (eds.) BPM 2016. LNCS, vol. 9850, pp. 73–89. Springer, Cham (2016). https://doi.org/10.1007/978-3-319-45348-4_5
5. Heiser, G.: Systems benchmarking crimes (2010). https://www.cse.unsw.edu.au/~gernot/benchmarking-crimes.html
6. van der Kouwe, E., Andriesse, D., Bos, H., Giuffrida, C., Heiser, G.: Benchmarking crimes: an emerging threat in systems security. arXiv preprint arXiv:1801.02381 (2018)
7. vanden Broucke, S.K.L.M., Delvaux, C., Freitas, J., Rogova, T., Vanthienen, J., Baesens, B.: Uncovering the relationship between event log characteristics and process discovery techniques. In: Lohmann, N., Song, M., Wohed, P. (eds.) BPM 2013. LNBIP, vol. 171, pp. 41–53. Springer, Cham (2014). https://doi.org/10.1007/978-3-319-06257-0_4
8. Tax, N., Lu, X., Sidorova, N., Fahland, D., van der Aalst, W.M.: The imprecisions of precision measures in process mining. Inf. Process. Lett. **135**, 1–8 (2018)
9. Janssenswillen, G., Donders, N., Jouck, T., Depaire, B.: A comparative study of existing quality measures for process discovery. Inf. Syst. **71**, 1–15 (2017)
10. Rozinat, A., de Medeiros, A.K.A., Günther, C.W., Weijters, A.J.M.M., van der Aalst, W.M.P.: The need for a process mining evaluation framework in research and practice. In: ter Hofstede, A., Benatallah, B., Paik, H.-Y. (eds.) BPM 2007. LNCS, vol. 4928, pp. 84–89. Springer, Heidelberg (2008). https://doi.org/10.1007/978-3-540-78238-4_10

11. Rozinat, A., De Medeiros, A.A., Günther, C.W., Weijters, A., Van der Aalst, W.M.: Towards an evaluation framework for process mining algorithms. Technical report 123, BPM Center Report (2007)
12. Weber, P., Bordbar, B., Tiňo, P., Majeed, B.: A framework for comparing process mining algorithms. In: GCC Conference and Exhibition, pp. 625–628. IEEE (2011)
13. vanden Broucke, S., De Weerdt, J., Vanthienen, J., Baesens, B.: A comprehensive benchmarking framework (CoBeFra) for conformance analysis between procedural process models and event logs in ProM. In: Proceedings of the IEEE Symposium on Computational Intelligence and Data Mining (CIDM 2013), pp. 254–261. IEEE (2013)
14. Depaire, B.: Process model realism: measuring implicit realism. In: Fournier, F., Mendling, J. (eds.) BPM 2014. LNBIP, vol. 202, pp. 342–352. Springer, Cham (2015). https://doi.org/10.1007/978-3-319-15895-2_29
15. Rehse, J.-R., Fettke, P., Loos, P.: Process mining and the black swan: an empirical analysis of the influence of unobserved behavior on the quality of mined process models. In: Teniente, E., Weidlich, M. (eds.) BPM 2017. LNBIP, vol. 308, pp. 256–268. Springer, Cham (2018). https://doi.org/10.1007/978-3-319-74030-0_19
16. Weijters, A., Ribeiro, J.T.S.: Flexible Heuristics Miner (FHM). In: IEEE Symposium on Computational Intelligence and Data Mining (CIDM), pp. 310–317. IEEE (2011)

A Logical Formalization of Time-Critical Processes with Resources

Carlo Combi$^{(\boxtimes)}$, Pietro Sala, and Francesca Zerbato

Department of Computer Science, University of Verona, Verona, Italy
{carlo.combi,pietro.sala,francesca.zerbato}@univr.it

Abstract. Checking time-critical properties of concurrent process instances having a finite amount of allocated resources is a challenging task. Modelling and understanding at design time the interactions of concurrent activities along the time line can become quite cumbersome, even for expert designers. In this paper, we consider processes that are composed of activities having a constrained duration and a bounded number of allocated resources, and we rely on a well-studied first order formalism, called $FO^2(\sim, <, -)$, to model and verify the interdependencies among multiple and concurrent process instances. Then, we show the expressiveness of our approach by describing the temporal properties that may be expressed through it. Throughout all the paper, we refer to a real clinical scenario to motivate our approach and showcase its expressiveness.

1 Introduction

In many application domains, such as manufacturing, emergency care, and logistics, it is desirable to achieve high quality organizational objectives while containing costs, reducing operating time, and limiting resource consumption [11]. Nevertheless, the efficient management and deadlock-free scheduling of limited resource amounts in time-critical scenarios remains a challenging task [25].

Processes consist of activities that are executed in a coordinated way to achieve a predefined goal [10] functional to organizational objectives. More specifically, process models are used to represent activities, their dependencies, and other aspects relevant for execution, such as temporal constraints, and criteria for resource allocation.

Time plays a crucial role in process design as understanding temporal properties and checking constraint satisfiability are crucial for process implementation. Indeed, violations of temporal constraints increase process costs and often lead to reduced service quality [11]. Time regulates multiple process aspects beside activity execution. Among them, that of scheduling and allocating resources remains one of the most challenging, and complexity increases when dealing with shared resource constraints [8]. In general, each process task is assigned to a resource of human or automatized type, entitled of executing it. However, the same process model acts as a blueprint for a set of instances, which have

© Springer Nature Switzerland AG 2018
M. Weske et al. (Eds.): BPM Forum 2018, LNBIP 329, pp. 20–36, 2018.
https://doi.org/10.1007/978-3-319-98651-7_2

to be temporally coordinated and often share the same resources [8,14]. When multiple process models are instantiated, each one possibly many times, instance coordination requires a lot of effort. Moreover, managing resources while observing temporal and allocation constraints can become challenging, even for expert process modelers. In such context, the main source of complexity stems from the need of achieving the final organizational objectives within the shortest time possible and with limited resource availability.

On the one hand, the time that each resource spends for executing the assigned tasks could be decreased only to a defined limit, as activities require a certain (constrained) amount of time to be executed. On the other hand, increasing workforce is critical for many organizations, especially in terms of costs containment, role sub-division, and overall resource coordination. As a result, processes often need to observe temporal and resource allocation constraints. In this setting, formal methods, such as first order and/or modal logical formalisms, can be exploited to find the trade-off between minimizing resource consumption and maximizing process outcomes, regardless of the scheduling criteria adopted for resource allocation and of the kinds of time constraints that are defined for a process [2,24].

In this work, we make use of a widely studied first order formalism $(FO^2(\sim, <, -))$ [4,19], for modeling and verifying processes having limited amount of allocated resources in time-critical scenarios. In particular, as an original contribution, we propose a strategy for detecting and resolving at design time shortcomings related to resource allocation in temporally constrained processes. The latter ones are specified as diagrams through the well-known Business Process Model and Notation (BPMN) [22]. To this end, we firstly retrieve information regarding the premises that may lead to resource unavailability. Then, depending on the addressed context, we may either decide to increase the number of allocated resources or propose alternative procedures for unraveling resource unavailability. In this work we will consider a real world scenario, taken from the clinical domain of emergency medicine, to motivate and discuss the proposed solution.

The remainder of the paper is organized as follows. Section 2 introduces related work. Section 3 presents the scenario taken from the domain of emergency medicine, used to show which kinds of (temporal) properties may be encoded with our approach. Section 4 explains the adopted temporal model. Section 5 introduces the logic $FO^2(\sim, <, -)$ and provides the mapping of process fragments into $FO^2(\sim, <, -)$. Section 6 discusses the proposed approach w.r.t. properties, complexity issues and expressiveness of $FO^2(\sim, <, -)$. Finally, Sect. 7 draws some conclusions.

2   Related Work

In this section, we discuss research studies related to shared-resource systems, conducted in the fields of business process management and theoretical logics.

When dealing with business process execution, it is desirable to perform processes effectively and efficiently with constrained time and resources [11].

The resource perspective of business processes deals both with the definition of assignments at design time, and with the allocation of resources at runtime. In [6] the authors propose a graphical notation for defining the assignments of human resources to business process activities. The novel notation, called RALph, allows one to define all the resource selection conditions specified by the workflow resource patterns [23], and supports their automatic analysis. As for dynamic process verification, in [16], the authors address the need of checking whether a temporal constraint defined on a single workflow being executed in concurrency with other workflows can be satisfied at a certain given time point. In the same direction, the method presented in [8] suggests to use sprouting graphs to detect and solve temporal violation paths in concurrent processes. In general, most challenges related to the specification and verification of resource allocation in BPMN processes derive from the limited expressiveness of *swimlanes*, that, for instance, do not allow designers to model separation and binding of duty constraints [14]. In this paper, we show how such limitations can be overcome by existing logic formalisms, which also allow the verification of resource allocation patterns in business processes. Verification of distributed systems or algorithms is a classical topic in theoretical computer science [18] and the need for a unified framework that enables both the specification and verification of distributed algorithms remains a hot topic in the field [1]. The satisfaction of constraints involving a combination of time durations and a limited amount of resources are typical compliance rule patterns in *monitoring the compliance of BPMN processes* (see [17] for a comprehensive survey of the topic). Our approach is intended for static verification but compliance rules may be expressed in our formalism as well.

The decidability status resulting from the extension of both first order [15] and modal [20] decidable logics with an equivalence relation has been widely studied. The usual applications of these logics are data-words [4,12]. In these research proposals, equivalence relations are used for comparing data located in different positions of the word. Our approach is somewhat orthogonal to those proposed in literature, as we consider processes having a bounded number of resources, yet allowing the amount of created process instances to be potentially unlimited. Such processes are implemented in a temporal model by using distinct classes of an equivalence relation, in order to guarantee that resources are assigned to process tasks in a mutually exclusive way.

3   A Motivating Clinical Example

Let us consider the management of patients in the emergency room (ER) of a hospital. The ER staff is formed by $\#nu$ nurses, denoted by variables $N_1, \ldots, N_{\#nu}$, $\#ph$ physicians, denoted by variables $P_1, \ldots, P_{\#ph}$, $\#pa$ teams of paramedics $A_1, \ldots, A_{\#pa}$ (each team is usually composed by several people including the driver), and a call center where $\#op$ operators $O_1, \ldots, O_{\#op}$ are responsible for answering emergency calls and for coordinating the initial communication between paramedics and ER clinicians ($\#name$ denotes the number of available resources of type *name* available in our system). For optimal patient management, the mentioned resources must act in coordination to perform a timely and multidisciplinary intervention on the patient. Resource availability must be monitored and process actors must be scheduled in order to guarantee that all incoming patients can be provided with proper care. Moreover, medical activities are subjected to temporal constraints, which limit the minimum and maximum duration allowed for their execution. As a motivating example, let us focus on the process of rescuing a severely injured patient who requires a blood transfusion. At a high level of abstraction, the macro-steps of patient care can be summarized as follows:

- an operator O_i receives an emergency call;
- a team of paramedics A_j is sent out to rescue patient *pat*;
- when patient *pat* is admitted into hospital, he or she is either directly examined by physician P_k or, if all the physicians P_k with $1 \leq k \leq \#ph$ are temporarily unavailable, *pat* is taken care of by nurse N_h, while waiting to be visited by some P_k;
- during examination, physician P_k assesses the severity of patient's conditions and determines how much blood is needed for treating patient *pat*.

The whole process is shown in Fig. 1, by using the standard Business Process Model and Notation [22]. BPMN allows us to represent processes in a graphical and understandable form,without compromising the generality of our approach. The diagram of Fig. 1, which describes the whole care process, is composed by three processes, each one enclosed within a *pool*. Pools are used to represent process resources, and can further be subdivided into *lanes*, to denote role specialization. In Fig. 1, one pool represents the team of paramedics, one is for the operator, and the rightmost one depicts clinicians in the ER and it is partitioned into two lanes, one for physicians and one for nurses. Communication between pools is based on message exchange: the same labels (and colors) are used to denote which are the corresponding messages exchanged between pools and a message flow is used to connect *send tasks* to the corresponding *receive events*. Messages are represented as *events*. In Fig. 1 *message flows* between corresponding send/receive events are omitted for readability reasons.

The process begins in the pool **Operator**. A new instance is created whenever an operator O_i receives an **Incoming emergency call**, as represented by the corresponding start message event. During the call, O_i must **Get informed about patient status and location**, such as a patient ID, his or her current position, and blood type. This is represented as a BPMN *task*, which is an atomic unit of work within the process, which is graphically depicted as a rounded box. However, in some circumstances, the blood type of patient *pat* remains unknown, as for instance, when *pat* is unconscious. Once the patient's position is known, the operator must **Request paramedics intervention** and **Alert clinicians**. These tasks are executed concurrently, as denoted by the preceding *parallel gateway*, marked with a "+". Both these actions trigger the corresponding processes in pools **Paramedics** and **Emergency Room**, as denoted by the depicted message flow. In particular, a team of paramedics A_j is sent to rescue the patient. Then, if the

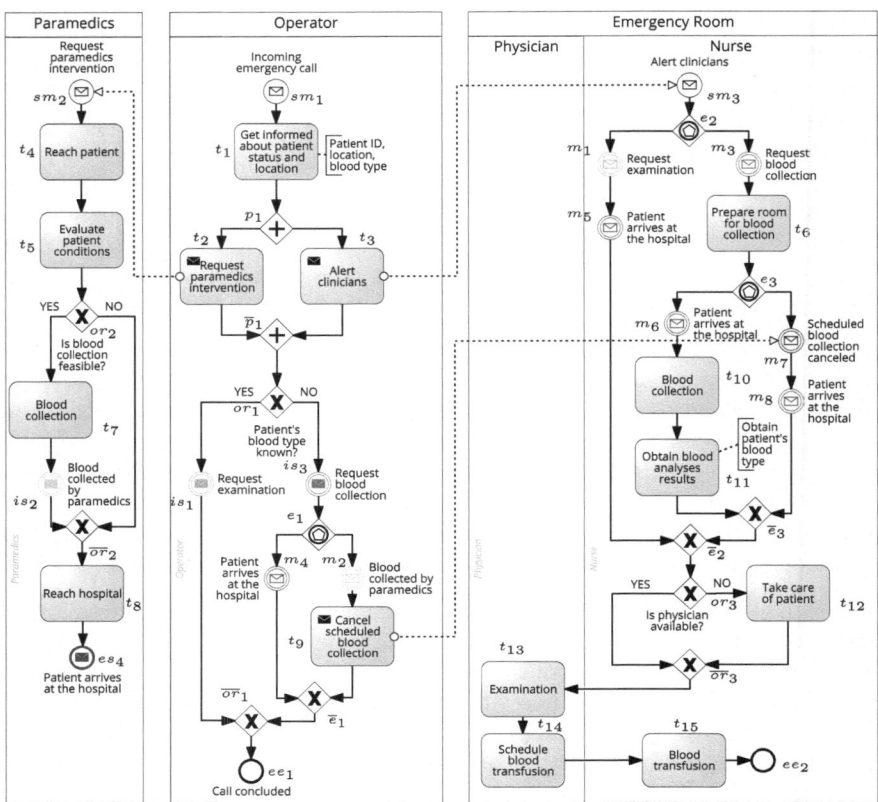

Fig. 1. BPMN process representing the main care actions performed to rescue and treat a severely injured patient who needs blood transfusion in ER.

patient's blood type is known, O_i must communicate with a nurse in the ER to Request Examination, or to Request blood collection otherwise, in order to guarantee that blood samples are drawn as soon as the patient is admitted into hospital. The choice between these two actions is represented by the *exclusive gateway* labeled Patient's blood type known? and marked with a "×". In the meanwhile, O_i remains in touch with the paramedics team A_j to monitor the transfer of the patient. If blood collection is feasible, paramedics collect it while transporting the patient to the hospital. In this case, O_i must communicate with a nurse to Cancel scheduled blood collection, as the patient can be directly examined. Once the patient arrives at the hospital, the paramedics process ends and, soon after, the operator's call is also concluded, thus O_i becomes available for receiving another emergency call. The remaining part of the ER process deals with hospital care. If the blood type of the patient is unknown, a nurse N_j performs a Blood collection and, then, must Obtain blood analyses results. Then, the patient is ready for Examination. However, if the physician is momentarily unavailable, a nurse must Take care of patient *pat* in the meanwhile. In this setting, it is easy to see how the mutual exclusion of resources dramatically affects the execution of a single process instance in presence of multiple instances that use the same resources. In Sect. 6, given a fixed amount of resources R, we will provide a way to compute the minimum execution time needed for completing P instances of the same process that use the resources in R. Finally, let us observe that the diagram of Fig. 1 is well-structured [9], from now on we assume all the diagrams to be well-structured.

4 A Temporal Model for Processes with Resources

In this section, we describe how to represent instances of processes such as those of Sect. 3 through a temporal model, which will be formalized by $FO^2(\sim, <, -)$ in Sect. 5.

Let $\mathbb{N}$ be our time domain. We begin by providing the semantics associated to points of the time domain (i.e., elements of $\mathbb{N}$). The *time frame partition* Tf of $\mathbb{N}$ with period δ is defined as the partition $\{[0, \ldots, \delta - 1][\delta, \ldots, 2\delta - 1] \ldots\}$ of $\mathbb{N}$ into consecutive intervals of length δ (i.e., $\mathit{Tf} = \{tf_k : k \in \mathbb{N}\}$ where $tf_k = [\delta k, \delta(k + 1) - 1]$), as shown in Fig. 2. Tf is the minimum temporal granularity of our domain and, without loss of generality, we assume that Tf is a temporal granularity of "minutes". All events happening within the same time frame tf_k are considered simultaneous. For instance, let us suppose that $\delta = 5$. According to the provided semantics, two events happening at time points 7 and 9 belong to the same time frame t_1 and, thus, we can assert that they both occur at minute 1.

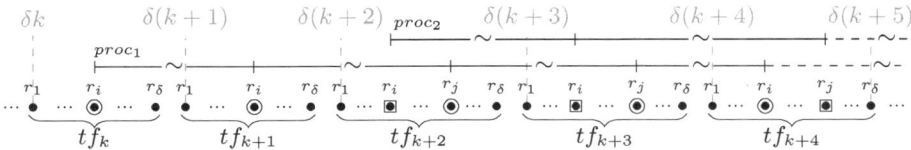

Fig. 2. A graphical account of how the time line is split into frames of size δ.

In general, an event happening at time point x occurs in the time frame tf_k, where $k = \lfloor \frac{x}{\delta} \rfloor$. For instance, if event ev_1 occurs in the time frame k and event ev_2 occurs in the time frame $k+h$, we have that ev_2 occurs exactly h minutes after e_1. Partitioning the time domain in this way allows us, given a point $x \in \mathbb{N}$, to claim that the behavior of the resource $(x\%\delta) + 1$ at the time $\lfloor \frac{x}{\delta} \rfloor$ (where function $\%$ denotes the remainder of the integer division) is represented through the specific features of point x. Thus, in our model, processes are represented by equivalence classes of an equivalence relation $\sim$ over $\mathbb{N}$. If two distinct natural numbers x and y satisfy $x \sim y$, then resource $r_{(x\%\delta)+1}$ and resource $r_{(y\%\delta)+1}$ are used by the same process at minutes $\lfloor \frac{x}{\delta} \rfloor$ and $\lfloor \frac{y}{\delta} \rfloor$, respectively. Let us consider the example depicted in Fig. 2. We have δ resources $r_1 \ldots r_\delta$ and two processes $proc_1$ and $proc_2$ represented as two different equivalence classes of some equivalence relation $\sim$. We have that $proc_1$ makes use of resource r_i at minutes k, $k+1$ and at minute $k+4$. This means that in such time frames $proc_2$ cannot use resource r_i since the equivalence relation prevents the same resource to be used by two distinct process in the same time frame. On the other hand, $proc_2$ uses resource r_i at minutes $k+2$ and $k+3$, while at minute $k+4$ it uses some other resource r_j.

As an example, Fig. 3 shows a (partial) execution of two instances of the BPMN process of Fig. 1, represented in our temporal model. Each process is related to a single patient and, overall, we have one operator, one team of paramedics, two nurses, and one physician. At each time frame tf_{k+h}, we have that each resource is associated to the element or fragment of the process that is being executed. If this is an activity, we assume that the resource is directly executing it, otherwise we assume that the resource is waiting for that part of the process to be executed by the engine. The two process-instances correspond to distinct equivalence classes and are represented by the doubly circled and the squared nodes respectively. It is worth noticing that every time frame contains exactly one occurrence for each resource. For such reason, a class that contains exactly one point denotes a resource that does not participate to any process instance for the specific time frame. For instance, the (unique) team of

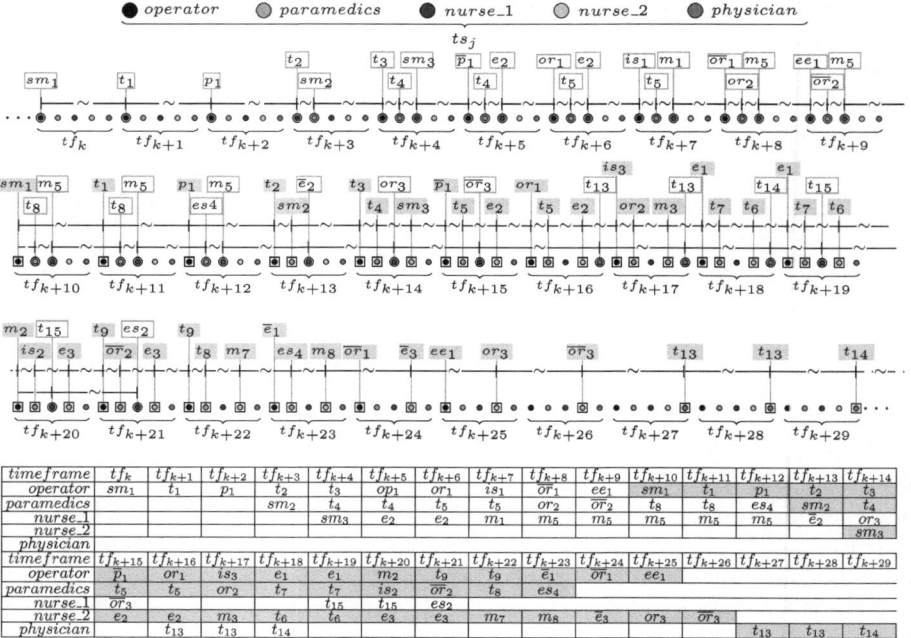

timeframe	tf_k	tf_{k+1}	tf_{k+2}	tf_{k+3}	tf_{k+4}	tf_{k+5}	tf_{k+6}	tf_{k+7}	tf_{k+8}	tf_{k+9}	tf_{k+10}	tf_{k+11}	tf_{k+12}	tf_{k+13}	tf_{k+14}
operator	sm_1	t_1	p_1	t_2	t_3	op_1	or_1	is_1	$\overline{or}_1$	ee_1	sm_1	t_1	p_1	t_2	t_3
paramedics				sm_2	t_4	t_4	t_5	t_5	or_2	$\overline{or}_2$	t_8	t_8	es_4	sm_2	t_4
nurse_1					sm_3	e_2	e_2	m_1	m_5	m_5	m_5	m_5	m_5	$\overline{e}_2$	or_3
nurse_2															sm_3
physician															
timeframe	tf_{k+15}	tf_{k+16}	tf_{k+17}	tf_{k+18}	tf_{k+19}	tf_{k+20}	tf_{k+21}	tf_{k+22}	tf_{k+23}	tf_{k+24}	tf_{k+25}	tf_{k+26}	tf_{k+27}	tf_{k+28}	tf_{k+29}
operator	$\overline{p}_1$	or_1	is_3	e_1	e_1	m_2	t_9	t_9	$\overline{e}_1$	$\overline{or}_1$	ee_1				
paramedics	t_5	t_5	or_2	t_7	t_7	is_2	$\overline{or}_2$	t_8	es_4						
nurse_1	$\overline{or}_3$			t_{15}	t_{15}	es_2									
nurse_2	e_2	e_2	m_3	t_6	t_6	e_3	e_3	m_7	m_8	$\overline{e}_3$	or_3	$\overline{or}_3$			
physician		t_{13}	t_{13}	t_{14}								t_{13}	t_{13}	t_{14}	

Fig. 3. Sample execution of the BPMN process of Fig. 1 considering two instances, represented both as a table and as a temporal model over $\mathbb{N}$.

paramedics is idle in the time frames going from tf_k to tf_{k+2} included and for the ones going from tf_{k+26} to tf_{k+29} included. We assume a semantics in which each traversed component consumes one time frame with the sole exceptions of tasks, which may take more than one unit of time, and of event-based gateways and receive-message components which, obviously, are waiting for some message to be produced. For instance, $nurse\_1$ at time frames tf_{k+5} and tf_{k+6} is waiting for either message m_1 or message m_3 and thus she stays in e_2 for these two time frames. At time frame tf_{k+7} the *operator*, by executing is_1 for the same patient/equivalence-class on which $nurse\_1$ was waiting, allows $nurse\_1$ to move to m_1 component. Finally, it is worth noticing that, given an equivalence class, the sequence of all its labelings represents a correct execution of the process for one patient.

5 Modeling Business Process Execution in $FO^2(\sim, <, -)$

In this section, we describe the steps to build a comprehensive $FO^2(\sim, <, -)$ formula that forces all its models to comply to the temporal specification of Sect. 4. First, we introduce the syntax and the semantics of $FO^2(\sim, <, -)$ and, then, we provide the mapping of BPMN diagrams into $FO^2(\sim, <, -)$ formulas.

The logic $FO^2(\sim, <, -)$ introduced below features a slightly different syntax w.r.t. the one presented in [4], namely $FO^2(\sim, <, +1)$. Yet, it is easy to see that the two logics have the same expressive power over $\mathbb{N}$. As previously mentioned, our domain consists of the set of natural numbers $\mathbb{N}$ and of a countable set Σ of unary relations σ. The syntax of $FO^2(\sim, <, -)$ is given by the following context-free grammar:

$$\varphi ::= \sigma(v) \mid v \sim v \mid v - v < n \mid \neg\varphi \mid \varphi \vee \varphi \mid \exists v\varphi, \ v ::= x \mid y, \text{ where } n \in \mathbb{N}, \ \sigma \in \Sigma.$$

The semantics of $FO^2(\sim, <, -)$ formulas is given through a mapping $\mathcal{M} : \Sigma \to 2^{\mathbb{N}}$, a function $f : \{x, y\} \to \mathbb{N}$, and an equivalence relation $\sim \subseteq \mathbb{N} \times \mathbb{N}$ as follows:

- $(\mathcal{M}, \sim, f) \models \sigma(v)$ if and only if $f(v) \in \mathcal{M}(\sigma)$;
- $(\mathcal{M}, \sim, f) \models v \sim v'$ if and only if $f(v) \sim f(v')$;
- $(\mathcal{M}, \sim, f) \models v - v' < n$ if and only if $f(v) - f(v') < n$;
- $(\mathcal{M}, \sim, f) \models \neg\varphi$ if and only if $(\mathcal{M}, \sim, f) \not\models \varphi$;
- $(\mathcal{M}, \sim, f) \models \varphi_1 \vee \varphi_2$ if and only if $(\mathcal{M}, \sim, f) \models \varphi_1$ or $(\mathcal{M}, \sim, f) \models \varphi_2$;
- $(\mathcal{M}, \sim, f) \models \exists v\varphi$ if and only if there exists $n \in \mathbb{N}$ such that $(\mathcal{M}, \sim, f') \models \varphi$ where $f'(v') = n$ if $v' = v$ and $f'(v') = f(v')$ otherwise.

In the following we will use the standard mathematical notation $[n]_\sim$ that denotes the equivalence class of n for every $n \in \mathbb{N}$ (i.e., $[n]_\sim = \{n' : n' \sim n\}$).

Given a BPMN diagram D and a set of resources $\mathcal{R}$ we provide a $FO^2(\sim, <, -)$ formula φ such that every possible execution of D using just resources $\mathcal{R}$ represents a model of φ and, on the other hand, every model of φ may be mapped into an execution of D using just resources $\mathcal{R}$. In the following, we define a few shorthands in order to make the mapping more compact and

$\forall v \varphi$	$\neg \exists v \neg \varphi$	$\varphi_1 \wedge \varphi_2$	$\neg(\neg\varphi_1 \vee \neg\varphi_2)$
$\varphi_1 \rightarrow \varphi_2$	$\neg\varphi_1 \vee \varphi_2$	$\varphi_1 \leftrightarrow \varphi_2$	$(\varphi_1 \rightarrow \varphi_2) \wedge (\varphi_2 \rightarrow \varphi_1)$
$v - v' = n$	$\neg(v - v' < n) \wedge \neg(v' - v < n)$	$|v - v'| < n$	$v - v' < n \wedge v' - v < n$
$v - v' > n$	$\neg(v - v' < n) \wedge \neg(v - v' = n)$	$n < v - v' < n'$	$n < v - v' \wedge v - v' < n'$
$v \simeq v'$	$|v - v'| < \delta \wedge (even(v) \leftrightarrow even(v'))$	$v \rightsquigarrow v'$	$0 < v' - v < 2\delta \wedge v' \sim v \wedge$ $(even(v') \leftrightarrow \neg even(v))$
$v < v'$	$v' - v > 0$	$v \precsim v'$	$v' > v \wedge v \sim v'$
$v \precsim_\delta v'$	$v \precsim v' \wedge v' - v = \delta$	$v \not\sim v'$	$\neg(v \sim v')$
$v \neq v'$	$\neg(v \simeq v')$	$v \leq v'$	$v < v' \vee v = v'$

Fig. 4. A set of useful shorthands for writing $FO^2(\sim, <, -)$ formulas.

readable, as reported in Fig. 4. Some encodings correspond to classical logical connectives, whereas some others are specific of our encoding (the *even* unary predicate may be forced to hold on $n \in \mathbb{N}$ if and only if n belongs to an even time-frame). In particular, for every pair of points x and y we have that $x \simeq y$ if and only if x and y belong to the same time frame, while $x \rightsquigarrow y$ if and only if y belongs to the time frame next to the time frame containing x. Before describing how each BPMN element or process fragment is translated into $FO^2(\sim, <, -)$, we show how to force a model to comply with the temporal model described in Sect. 4. Let $\mathcal{R} = \{R_1 \ldots R_\delta\}$ be the set of resources, which we assume being also unary relations. First, we impose that every point $x \in \mathbb{N}$ is labeled with exactly one resource, which is repeated for all points $x + k\delta$ with $k \in \mathbb{N}$. This is done by means of the following formula:

$$\bigwedge_{R \in \mathcal{R}} \exists x R(x) \wedge \forall x (\bigvee_{R \in \mathcal{R}} R(x)) \wedge \forall x (\bigwedge_{R \in \mathcal{R}} (R(x) \rightarrow \bigwedge_{R' \in \mathcal{R} \setminus \{R\}} \neg R'(x)) \wedge \bigwedge_{R \in \mathcal{R}} (R(x) \rightarrow \exists y (R(y) \wedge y - x = \delta)).$$

Moreover, let $\mathcal{T}$ be the set of tasks in our BPMN diagrams (e.g., in Fig. 1 we have $\mathcal{T} = \{t_1, \ldots, t_{15}\}$), tasks are represented by unary variables in our model (i.e., $\mathcal{T} \subseteq \Sigma$). We impose that a resource may execute at most one task in a single time frame by means of the formula: $\forall x (\bigvee_{t \in \mathcal{T}} (t(x) \rightarrow \bigwedge_{t' \in \mathcal{T} \setminus \{t\}} \neg t'(x)))$.

For our mapping we take a BPMN diagram as input and we partition the set of resources $\mathcal{R}$ into roles represented in BPMN through lanes. For instance, in the described motivating scenario, we have that $\mathcal{R}$ is partitioned in $Operator = \{O_1, \ldots, O_{\#op}\}, Paramedic = \{A_1, \ldots, A_{\#pa}\}, Nurse = \{N_1, \ldots, N_{\#nu}\}$, and $Physician = \{P_1, \ldots, P_{\#ph}\}$. For each role $Role$ we define the corresponding formula as $Role(x) = \bigvee_{R \in Role} R(x)$ and thus $\delta = \#op + \#pa + \#nu + \#ph$. In our example, we have that $Nurse(x) = \bigvee_{N_i, 1 \leq i \leq \#nu} N_i(x)$. In the proposed mapping, each BPMN fragment c of the process diagram is uniquely identified by a unary relation. Thus, for every resource $R \in \mathcal{R}$, every BPMN fragment c in the input diagram, and every $n \in \mathbb{N}$, if both $R(n)$ and $c(n)$ hold, then resource R in the time frame $\lfloor \frac{n}{\delta} \rfloor$ is engaged in executing fragment c. In our mapping, we impose that a resource R executing a component c is blocked until the execution of c terminates. More precisely, for every point n for which both $R(n)$ and $c(n)$ hold, if there exists a point $n' \sim n$ with $\lfloor \frac{n'}{\delta} \rfloor = \lfloor \frac{n}{\delta} \rfloor + 1$ where $c(n')$ holds (i.e., c has not finished its execution at $\lfloor \frac{n'}{\delta} \rfloor$ for the process instance $[n]_\sim$), then $R(n')$ must hold. For instance, in the example of Sect. 4 we have that, if a blood transfusion (task t_{15}) is assigned to a nurse N_j in some process $[n]_\sim$, for all the following time frames, all the points in $t_{15} \cap [n]_\sim$ will belong to N_j as well, until the transfusion finishes.

We may want to specify further, more specific, constraints such as binding of duties [6]. In our example, this constraint regards paramedics, since for every patient $[n]_\sim$ we have that the team of paramedics that reaches $[n]_\sim$ and the one that take him to the hospital must be the same. We call this property *lane-level atomicity*, which is a coarser atomicity than the component-level one. Given a *Role* lane, in our mapping we force only the component-level atomicity, but we can force lane-level atomicity for just the roles on which it is required by means of the formula $\forall x(\bigwedge_{R \in Role}(R(x) \rightarrow \forall y(y \sim x \rightarrow R(y))))$. We may be even more precise. For instance, we may require that, for a given patient the nurse that transfuses blood (t_{15}) must be the one who previously collected the blood (t_{10}). In our example of Fig. 3 this is done by means of the formula: $\forall x(t_{10}(x) \wedge \exists y(x \sim y \wedge t_{15}(y)) \rightarrow \bigwedge_{i \in 1,2}(N_i(x) \rightarrow \exists y(x \sim y \wedge t_{15}(y) \wedge N_i(y))))$.

In Fig. 5, we provide the encoding for BPMN tasks and message events, while in Fig. 6 we provide the mapping for process fragments enclosing gateways. For space reasons, the presented encoding refers to fragments of the process in Fig. 1, but the mapping may be easily extended to the complete BPMN notation.

The full encoding of D is the conjunction of all the $FO^2(\sim, <, -)$ formulas that encode each process fragment. As an example, let us consider the mapping of an intermediate catching message event. We distinguish here two cases, which are encoded in two different ways. The first case consists of m waiting for a message that is generated by some task t. The latter case consists of event m that is waiting for a message generated by one or more intermediate (resp. *end*) throwing events *is* (resp. *es*). Since in this case the message that m is waiting for may be generated by more than one intermediate (resp. *end*) throwing event *is* (resp. *es*), we identify the *sources of m* as the set $\mathcal{S}_m$ of all and only the related intermediate/end throwing message events. For instance, if we consider the diagram of Fig. 1 we have that $\mathcal{S}_{m_1} = \{is_1\}$ while $\mathcal{S}_{m_4} = \mathcal{S}_{m_5} = \mathcal{S}_{m_6} = \mathcal{S}_{m_8} = \{es_4\}$. For identifying points associated with a resource producing the message that m is waiting for we use formula $\mathcal{S}_m(v) = \bigvee_{c \in \mathcal{S}_m} c(v)$.

Figure 6 shows how to translate BPMN gateways. Every BPMN gateway may be either a split or a join. Since we assume D to be well-structured, we have that every split gateway c has a matching join gateway $\bar{c}$. Fig. 6 depicts the encodings for the split event-based, parallel, and exclusive gateways, while only the most interesting join gateways are reported. Let us consider now the split event-based gateway e. The difference between such split gateway and the other two is that such gateway may be forced to wait for the occurrence of one among its children catching events. The encoding of e makes use of a *trigger formula* θ_c for every catching event c. Such formulas are used to determine if the execution must leave e or must stay in it in the next time frame. In the considered BPMN fragment, one of these catching events c may be catched either by an intermediate/end

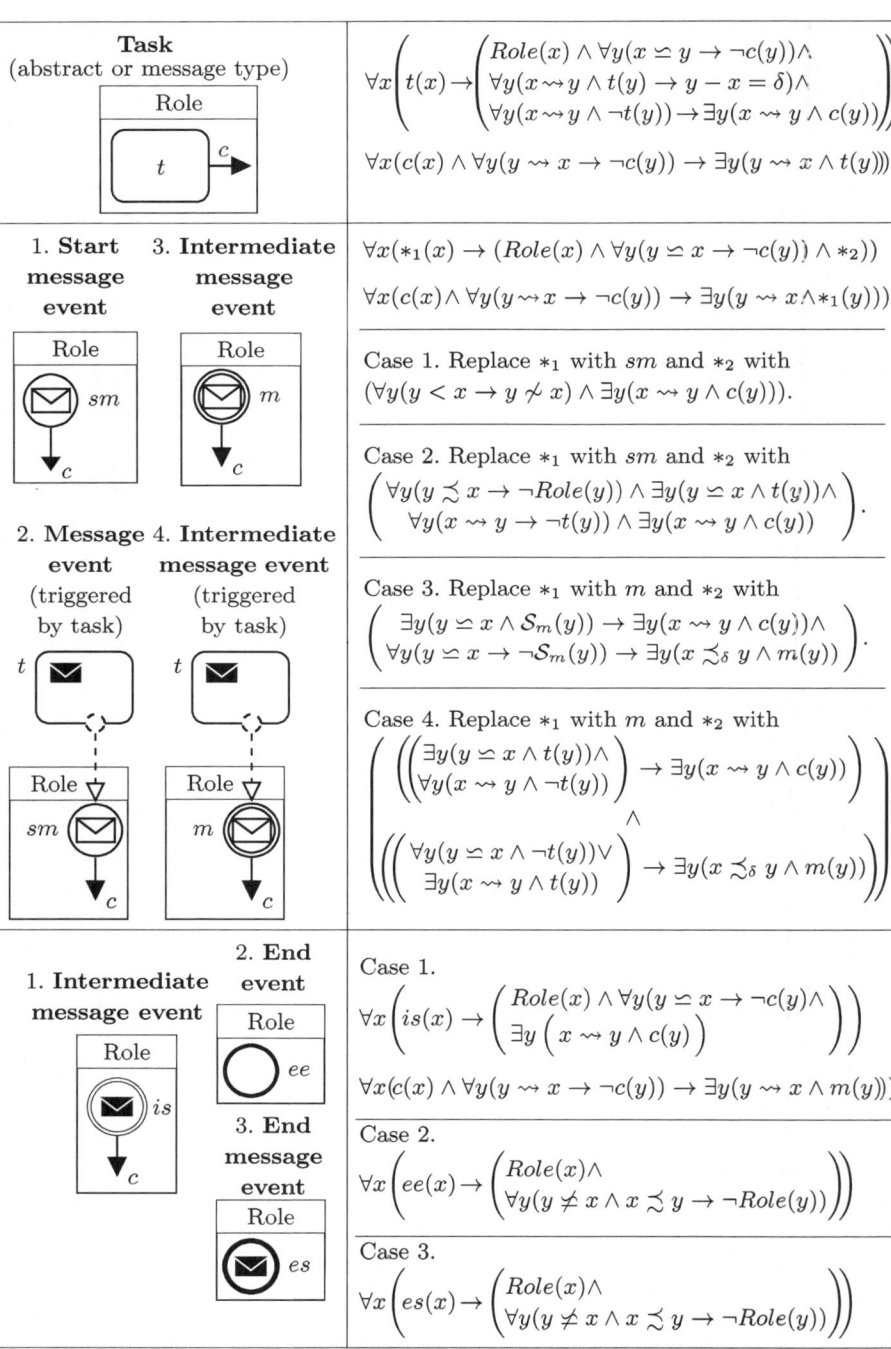

Fig. 5. Mapping of BPMN fragments containing tasks and message events.

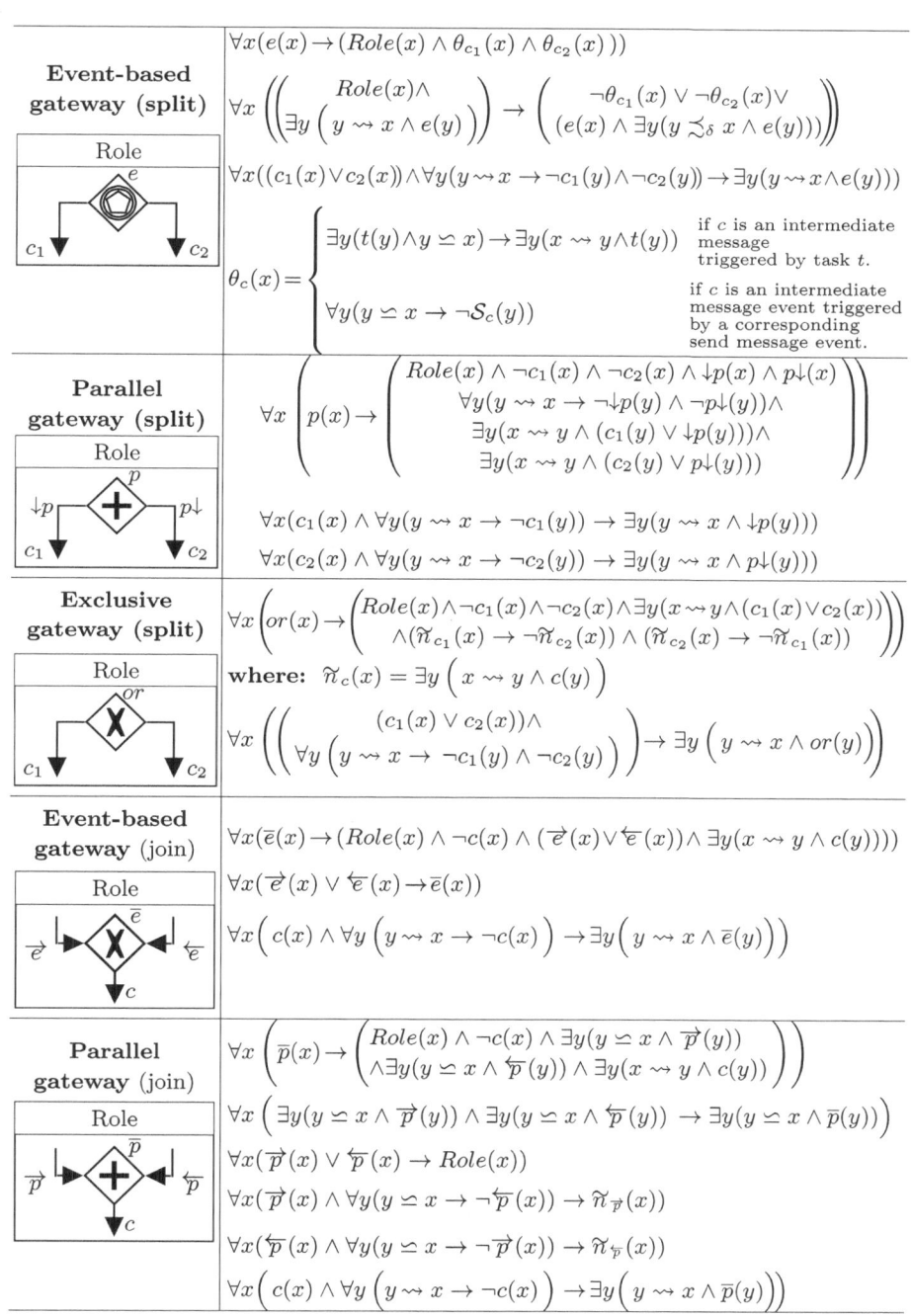

Fig. 6. Mapping into $FO^2(\sim, <, -)$ of BPMN fragments containing gateways.

send message component or by a task. As we may see in Fig. 6, θ_c depends on the type of the event c. Let us consider gateway e_3 in Fig. 1. Such gateway has two connected events, namely m_6 and m_7. In this case m_6 is triggered by event es_4 (i.e., $\mathcal{S}_{m_6} = \{es_4\}$) and m_7 is triggered by task t_9. Then we have $\theta_{m_6} = \forall y(y \simeq x \rightarrow \neg es_4(y))$ and $\theta_{m_7} = \exists y(t_9(y) \wedge y \simeq x) \rightarrow \exists y(x \rightsquigarrow y \wedge t_9(y))$. Gateway e_3 is translated as follows:

$\forall x(e_3(x) \rightarrow (Nurse(x) \wedge \theta_{m_6}(x) \wedge \theta_{m_7}(x)))$
$\forall x(Nurse(x) \wedge \exists y(y \rightsquigarrow x \wedge e_3(y)) \rightarrow \neg\theta_{m_6}(x) \vee \neg\theta_{m_7}(x) \vee (e_3(x) \wedge \exists y(y \precsim_\delta x \wedge e_3(y))))$
$\forall x((m_6(x) \vee m_7(x)) \wedge \forall y(y \rightsquigarrow x \rightarrow \neg m_6(y) \wedge \neg m_7(y)) \rightarrow \exists y(y \rightsquigarrow x \wedge e_3(y)))$

Now, let us focus on the join gateways. For every join $\bar{c}$ we introduce two auxiliary variables $\overrightarrow{c}$ and $\overleftarrow{c}$. In principle, such variables are not necessary but their use improves the readability and relieves us from managing tedious special cases. Variables $\overrightarrow{c}$ and $\overleftarrow{c}$ are used for the successor element for the left incoming element and the right incoming element of $\bar{c}$ respectively. For instance, in Fig. 1 the successor of t_2 is $\overrightarrow{p}_1$ and the successor of t_3 is $\overleftarrow{p}_1$ and thus $\overrightarrow{p}_1$ is the successor c in the encoding of t_2 and $\overleftarrow{p}_1$ is the successor c in the encoding of t_3. In the context of join gateways the encoding related to event-based and exclusive gateways is straightforward, while the join parallel gateway is a little bit more complicated since, in order to move on, it must wait for both the two incoming edges to terminate. For such reason, for each parallel split gateway we introduce two more auxiliary variables, namely $\downarrow p$ and $p \downarrow$, used for forcing the execution of both process "branches".

6   Properties, Complexity and Expressiveness

In this section, we detail some properties that may be checked/imposed through our formalism, then we discuss the complexity of the satisfiability problem for $FO^2(\sim, <, -)$ and, finally, we compare the expressiveness of our approach to similar ones present in the literature.

Qualitative Properties. We begin by showing a couple of a qualitative requirements that are desirable but not enforced by the BPMN process of Fig. 1. Constraints on the durations of tasks may be easily imposed. For instance, if we want a given task t to last more than k time points we just add the following constraint $\forall x(t(x) \wedge \neg\exists y(t(y) \wedge y \precsim_\delta x) \rightarrow \forall y(\bigvee_{i=1}^{k} y - x = i\delta \rightarrow x \sim y \wedge t(y)))$. By means of negation, conjunction and disjunction we may express common constraint like imposing that the duration of a task must be within a given interval and so on. Moreover, we may add constraints on the behaviour of the process based on resource availability. Let us consider, for instance, the Exclusive Gateway or_3 in Fig. 1 in a process execution it is perfectly licit that a physician resource is available when the execution arrives at or_3 but the the NO alternative is still taken and task t_{12} is performed. This situation may be easily avoided by means of the following formula $\neg\exists x(or_3(x) \wedge \exists y(x \rightsquigarrow y \wedge t_{12}(y)) \wedge \exists y(Physician(y) \wedge y \simeq x \wedge \forall x(x \precsim_\delta y \vee y \precsim_\delta x \rightarrow x \not\sim y)))$.

Quantitative Properties. Now we show how to check a couple of properties regarding the time needed for completing scenarios in presence of multiple instances of the same process sharing the same set of resources $\mathcal{R}$. Such properties may be easily checked using $FO^2(\sim, <, -)$. The first property we address is the following:

Property 1. Given a BPMN diagram D, a set of resources $\mathcal{R}$ and two natural numbers number k and n. There exists an execution of k instances of D that terminates before time frame tf_n using only the resources in $\mathcal{R}$.

Such property may be checked in $FO^2(\sim, <, -)$ as follows. Let $\psi(D, \mathcal{R})$ be the formula encoding the all the possible executions of multiple instances of D with resources in $\mathcal{R}$ that is obtained from the mapping proposed in Sect. 5. Let us assume w.l.o.g that sc is the element in D that denotes the beginning of the process (e.g., sm_1 in the diagram of Fig. 1). Let $pr_1, \ldots, pr_k$ a set of fresh unary variables, the formula that tells whether or not Property 1 holds for the input tuple $(D, \mathcal{R}, k, n)$ is obtained by the conjunction of $\psi(D, \mathcal{R})$ and the following formula $\psi(sc, \mathcal{R}, k, n)$ (let $n_{\mathcal{R}} = |\mathcal{R}| \cdot n$):

$$\bigwedge_{i=1}^{k} (\exists x (pr_i(x) \wedge sc(x))) \wedge \forall x \left(\bigwedge_{i=1}^{k} (pr_i(x) \rightarrow \forall y (y \sim x \rightarrow pr_i(y)) \wedge \bigwedge_{j=1, j \neq i}^{k} \neg pr_j(x)) \right)$$

$$\wedge \forall x \forall y (\bigvee_{i=1}^{k} pr_i(x) \rightarrow x - y \leq n_{\mathcal{R}})$$

The first two conjuncts imposes that there exists at least k distinct instances of the process, while the last imposes that all those instances terminates within the first n time frames. The second property we want to describe in $FO^2(\sim, <, -)$ is the following:

Property 2. Given a BPMN diagram D, a set of resources $\mathcal{R}$ and four natural numbers $k' < k$ and $n' < n$. There exists and execution that satisfies Property 1 on $(D, \mathcal{R}, k, n)$ for which at least k' processes terminate within time frame n'

This property may be checked, using $\psi(D, \mathcal{R})$ and $\psi(sc, \mathcal{R}, k, n)$, by means of the following formula (let $n'_{\mathcal{R}} = |\mathcal{R}| \cdot n'$):

$$\psi(D, \mathcal{R}) \wedge \psi(sc, \mathcal{R}, k, n) \wedge \exists x (\forall y (y < x \rightarrow x - y \leq n'_{\mathcal{R}}) \wedge \forall y (y > x \rightarrow \bigwedge_{j=1}^{k'} \neg pr_j(y)))$$

Having these two parametrized formulas we may apply a dichotomic search to solve the following two problems:

Problem 1. Given a BPMN diagram D, a set of resources $\mathcal{R}$ and a natural number k. What is the minimum n for which k instances of D that terminates before time frame tf_n using only the resources in $\mathcal{R}$.

Problem 2. Given a BPMN diagram D, a set of resources $\mathcal{R}$ and two natural numbers k and n. What are the Pareto optimal pairs (k', n') for which Property 2 holds on $(D, \mathcal{R}, k', k, n', n)$[1].

---

[1] A Pareto optimal solution is a solution that is not dominated by any other solution, in this case we are looking for pairs (k', n') with $k' < k$ and $n' < n$ such that Property 2 does not hold on $(D, \mathcal{R}, k'+1, k, n', n)$ and $(D, \mathcal{R}, k', k, n'+1, n)$.

Complexity. In general, the complexity of the satisfiability problem for the full logic $FO^2(\sim, <, -)$ is not known. As a matter of fact such problem, in the finite case, is equivalent to the reachability problem for Petri Nest/Vector Addition Systems [4]. For the infinite case the problem remains decidable with at least the same complexity of the finite one. This means that the best known algorithm for solving it turns out to be not primitive recursive. However, it is easy to prove that the mapping proposed in Sect. 5 forces the following property on $\sim$.

Definition 1. *Given an equivalence relation $\sim$ over $\mathbb{N}$ and a number $k \in \mathbb{N}$, we say that $\sim$ is k-pulsating if and only if for every $n \in \mathbb{N}$ either $n = \max([n]_\sim)$ or there exists $n' > n$ such that $n' \sim n$ and $n' - n \leq k$.*

Informally speaking, $\sim$ is k-pulsating if and only if for every point n which is not the maximum of its class there exists a point $n' \sim n$ greater than n but less or equal than $n + k$. It is worth noticing that a class $[n]_\sim$ may feature (not consecutive) points at an arbitrary distance as long as it keeps "pulsing". Due to the lack of space we cannot provide further details on how this property allows us to reduce the complexity of $FO^2(\sim, <, -)$ formulas that stem from the mapping proposed in Sect. 5. The complete result will be provided in the extended version of this work. In detail, we will prove that the satisfiability problem for such formulas is satisfiable only on models where $\sim$ is $|\mathcal{R}|$-pulsating. Finally, this result allows us to prove that the satisfiability problem for $FO^2(\sim, <, -)$ formulas originated by the mapping of Sect. 5 belongs to the complexity class PSPACE (resp., EXPSPACE) if $|\mathcal{R}|$ is represented by a unary (resp., binary) encoding.

Expressiveness. Digging deep into the literature for static BPMN verification most proposals make use of timed automata (e.g., [21]). Such approaches do not take into account resources allocation, which is instead a central goal in our proposal. Many tools for timed automata (e.g., [3]) allow to verify properties by means of (fragments of) the well known temporal logics LTL and CTL. Since CTL is a logic that works on branching structures, our comparison regards LTL which works on linear structures like $FO^2(\sim, <, -)$. $FO^2(\sim, <, -)$ may express all the operators of LTL but the unbounded until operator. As for the bounded until operator metric constraints may be used to express it [5]. On the other hand, in LTL (as well as in CTL) it is not possible to express an equivalence relation over points.

Interesting recent developments in verification of complex systems focus on the integration of processes and data [13] and are close enough to our approach since resources may be seen as a particular kind of data. Many of these works adopt a formalism based on transition systems where each state contains a first order representation of the data on a (possibly) infinite domain. Such systems are verified by using the product of the classical first-order formalism for dealing with the data content in each state, and some temporal logic such as LTL or μ-calculus, for dealing with transitions between states. The major drawback in this approach is that many of such logics turn out to be undecidable in the general case and thus sintactic and/or semantics restrictions are introduced in order to achieve decidability [7].

7   Conclusions

In this paper, we explored the challenges related to the specification and verification of concurrently running processes, operating in time-critical scenarios and having assigned a limited amount of resources. In particular, we proposed the use of the fragment of first order logic $FO^2(\sim, <, -)$ to capture process fragments along the timeline and to combine them in a sound model, by observing constraints defined on both activity durations and resource availabilities. As a motivating application domain, we introduced and discussed a real world ER scenario, involving three different and intertwined processes. Specifically, we provide a mapping from BPMN to $FO^2(\sim, <, -)$. Moreover, we addressed the formal verification of some time-related constraints. Finally, we addressed (briefly) the complexity of checking the satisfiability of $FO^2(\sim, <, -)$ formulas that represent BPMN processes and properties over them. In order to prove the feasibility of our approach, we plan to build a prototype that implements our mapping of BPMN diagrams in $FO^2(\sim, <, -)$ and verifies properties over it, to the best of our knowledge no direct prover for $FO^2(\sim, <, -)/FO^2(\sim, <, +1)$ has been developed yet.

Acknowledgments. The authors would like to thank the anonymous reviewers for their constructive criticism and the series of invaluable suggestions that will fuel the future developments of the present work.

References

1. Aiswarya, C., Bollig, B., Gastin, P.: An automata-theoretic approach to the verification of distributed algorithms. In: CONCUR 2015, vol. 42, pp. 340–353. Leibniz-Zentrum für Informatik (2015)
2. Attie, P.C., Singh, M.P., Emerson, E.A., Sheth, A., Rusinkiewicz, M.: Scheduling workflows by enforcing intertask dependencies. Distrib. Syst. Engi. **3**(4), 222–238 (1996)
3. Bengtsson, J., Larsen, K., Larsson, F., Pettersson, P., Yi, W.: UPPAAL — a tool suite for automatic verification of real-time systems. In: Alur, R., Henzinger, T.A., Sontag, E.D. (eds.) HS 1995. LNCS, vol. 1066, pp. 232–243. Springer, Heidelberg (1996). https://doi.org/10.1007/BFb0020949
4. Bojańczyk, M., David, C., Muscholl, A., Schwentick, T., Segoufin, L.: Two-variable logic on data words. ACM Trans. Comput. Log. **12**(4), 27:1–27:26 (2011)
5. Bresolin, D., Montanari, A., Sala, P., Sciavicco, G.: Optimal decision procedures for MPNL over finite structures, the natural numbers, and the integers. Theor. Comput. Sci. **493**, 98–115 (2013). https://doi.org/10.1016/j.tcs.2012.10.043
6. Cabanillas, C., Knuplesch, D., Resinas, M., Reichert, M., Mendling, J., Ruiz-Cortés, A.: RALph: a graphical notation for resource assignments in business processes. In: Zdravkovic, J., Kirikova, M., Johannesson, P. (eds.) CAiSE 2015. LNCS, vol. 9097, pp. 53–68. Springer, Cham (2015). https://doi.org/10.1007/978-3-319-19069-3_4
7. Calvanese, D., De Giacomo, G., Montali, M., Patrizi, F.: First-order μ-calculus over generic transition systems and applications to the situation calculus. Inf. Comput. **259**(3), 328–347 (2018). https://doi.org/10.1016/j.ic.2017.08.007

8. Du, Y., Xiong, P., Fan, Y., Li, X.: Dynamic checking and solution to temporal violations in concurrent workflow processes. IEEE Trans. Syst. Man Cybern. Part A Syst. Hum. **41**(6), 1166–1181 (2011)

9. Dumas, M., García-Bañuelos, L., Polyvyanyy, A.: Unraveling unstructured process models. In: Mendling, J., Weidlich, M., Weske, M. (eds.) BPMN 2010. LNBIP, vol. 67, pp. 1–7. Springer, Heidelberg (2010). https://doi.org/10.1007/978-3-642-16298-5_1

10. Dumas, M., La Rosa, M., Mendling, J., Reijers, H.A.: Fundamentals of Business Process Management, vol. 1. Springer, Heidelberg (2013). https://doi.org/10.1007/978-3-662-56509-4

11. Eder, J., Panagos, E., Rabinovich, M.: Workflow time management revisited. In: Bubenko, J., Krogstie, J., Pastor, O., Pernici, B., Rolland, C., Sølvberg, A. (eds.) Seminal Contributions to Information Systems Engineering, pp. 207–213. Springer, Heidelberg (2013). https://doi.org/10.1007/978-3-642-36926-1_16

12. Figueira, D.: A decidable two-way logic on data words. In: ACM/IEEE Symposium in Logics in Computer Science (LICS), pp. 365–374 (2011)

13. Hariri, B.B., Calvanese, D., De Giacomo, G., Deutsch, A., Montali, M.: Verification of relational data-centric dynamic systems with external services. PODS **2013**, 163–174 (2013)

14. Havur, G., Cabanillas, C., Mendling, J., Polleres, A.: Resource allocation with dependencies in business process management systems. In: La Rosa, M., Loos, P., Pastor, O. (eds.) BPM 2016. LNBIP, vol. 260, pp. 3–19. Springer, Cham (2016). https://doi.org/10.1007/978-3-319-45468-9_1

15. Kieronski, E., Tendera, L.: On finite satisfiability of two-variable first-order logic with equivalence relations. LICS **2009**, 123–132 (2009)

16. Li, H., Yang, Y.: Dynamic checking of temporal constraints for concurrent workflows. Electron. Commer. Res. Appl. **4**(2), 124–142 (2005)

17. Ly, L.T., Maggi, F.M., Montali, M., Rinderle-Ma, S., van der Aalst, W.M.P.: Compliance monitoring in business processes: functionalities, application, and tool-support. Inf. Syst. **54**, 209–234 (2015). https://doi.org/10.1016/j.is.2015.02.007

18. Manna, Z., Pnueli, A.: The Temporal Logic of Reactive and Concurrent Systems. Springer, New York (1992). https://doi.org/10.1007/978-1-4612-0931-7

19. Montanari, A., Pazzaglia, M., Sala, P.: Metric propositional neighborhood logic with an equivalence relation. Acta Inf. **53**(6–8), 621–648 (2016)

20. Montanari, A., Sala, P.: Interval-based synthesis. GandALF **2014**, 102–115 (2014)

21. Morales, L.E.M., Monsalve, C., Villavicencio, M.: Formal verification of business processes as timed automata. CISTI **2017**, 1–6 (2017)

22. Object Management Group: Business Process Model and Notation (BPMN), v2.0.2. www.omg.org/spec/BPMN/2.0.2/

23. Russell, N., van der Aalst, W.M.P., ter Hofstede, A.H.M., Edmond, D.: Workflow resource patterns: identification, representation and tool support. In: Pastor, O., Falcão e Cunha, J. (eds.) CAiSE 2005. LNCS, vol. 3520, pp. 216–232. Springer, Heidelberg (2005). https://doi.org/10.1007/11431855_16

24. Senkul, P., Toroslu, I.H.: An architecture for workflow scheduling under resource allocation constraints. Inf. Syst. **30**(5), 399–422 (2005)

25. Zhou, M., Fanti, M.P.: Deadlock Resolution in Computer-Integrated Systems. CRC Press, Boca Raton (2004)

Business Process and Rule Integration Approaches - An Empirical Analysis

Tianwa Chen[1], Wei Wang[1(✉)], Marta Indulska[2], and Shazia Sadiq[1]

[1] School of Information Technology and Electrical Engineering,
The University of Queensland, Brisbane, Australia
{tianwa.chen,w.wang9}@uq.edu.au,
shazia@itee.uq.edu.au
[2] University of Queensland Business School, The University of Queensland,
Brisbane, Australia
m.indulska@business.uq.edu.au

Abstract. Modeling landscapes in organizations suffer from the problem of information silos, where a number of process models, business rule repositories and other information artefacts may exist concurrently for the same business activity. In this paper, we investigate integrating business process models and business rules. Prior literature presents three such approaches, namely text annotation, diagrammatic, and link integration. We evaluate these approaches from a cognitive load perspective and measure the value of integration from three perspectives: understanding accuracy, mental effort and time efficiency. Our results indicate that diagrammatic integration is associated with better understanding accuracy than text annotation and link integration, but may require more mental effort and time under certain conditions. We also found that the integration approach partially influences mental effort and time efficiency. Further insights from our empirical analysis reveal relationships between process model constructs, integration approaches and cognitive load, especially how approaches applied to models with specific characteristics, impact on process understanding and cognitive load.

Keywords: Business rules · Integrated modeling · Business process modeling
Cognitive load

1 Introduction

Since the inception of business process modeling, the dual need of human understanding and executability of process models has been under discussion. Numerous studies have been conducted to gain insights into how these, often opposing, needs can be met. In practice, the understanding of a business process often depends on two aspects, that is, the understanding of the business process model and the understanding of any related business rules, which may or may not be part of the process model [29, 34, 36]. The understanding extracted from graphical process models is focused on the temporal or logical relationships between business activities, whereas the business rules comprise the constraints and mandates to control the behaviour of the business process and its activities [39]. When the two are not integrated, it increases the risks of

© Springer Nature Switzerland AG 2018
M. Weske et al. (Eds.): BPM Forum 2018, LNBIP 329, pp. 37–52, 2018.
https://doi.org/10.1007/978-3-319-98651-7_3

incomplete understanding of the business process and hampers the effectiveness of business process management.

To facilitate better understanding efficiency and effectiveness, several studies have advocated integration of business rules into business process model [3, 14, 15, 34, 36]. At the same time, however, there is evidence that existing business process modeling languages lack the representational capacity to represent business rules sufficiently [8, 27]. Due to such representational limitations of graphical process modeling techniques it is not always possible, or indeed desirable, to represent related business rules within the process model [27]. Several studies have also explored situations under which business process models and business rules are best kept separated, and those when they are best integrated [3, 7, 13, 36].

Prior research classified integration of business rules with business process models into three approaches, namely text annotation integration, diagrammatic integration and linked integration (see Fig. 1). Text integration is a way of representing business rules in business process models by adding textual descriptions of rules – e.g. in BPMN, using the BPMN text annotation construct [3, 7, 36]. In contrast, diagrammatic integration relies on control flow constructs, such as sequence and gateways, and other constructs to represent business rules in business process models [13, 36]. Linked integration is characterised by the use of external rule repository links. It can either use static or dynamic approach to integrate and link each business rule with the corresponding part of the business process model [32, 36].

Despite several studies proposing various approaches for business process and rule integration, there is limited knowledge on the effect these approaches have on process understanding [35]. Previous studies have demonstrated that the linked integration approach is associated with better business process understanding as compared to a separated representation of process model and related rules [36]. However, how the three different approaches to business process and rule integration compare in terms of process understanding remains unknown.

In this paper, we present the outcomes of an empirical analysis undertaken to study the effects of different process and rule integration approaches on business process model understanding. Using a cognitive load perspective, and with the help of eye tracking equipment, we conduct an experiment to compare the differences of link integration, text annotation and diagrammatic integration on business process model understanding. The experiment uses three measurements to conduct the comparison, namely, understanding accuracy, mental effort and time efficiency. Our study provides empirical findings on the relative merits of integration approaches, which can help modelers make informed decisions regarding integration of rules and process models.

In the following sections we first present the research background of business rule integration methods as well as the role of eye tracking methods in studying business process model understanding. Section 3 introduces our experiment design. Section 4 presents the data analysis methods, the results of the experiment and discussion of insights drawn from the results, and finally Sect. 5 summarizes the contribution of the paper, limitations of the study, and an outline of future extensions of this work.

2   Related Work

Business process modeling and business rule modeling are complementary approaches for modeling business activities. To improve business process model representational capacity, researchers have developed various business rules integration methods in literature [14, 15]. In summary, three approaches of business rules integration methods have been proposed, namely, text annotation, diagrammatic integration and link integration, as shown in Fig. 1. It can be observed that the three integration methods have various distinctions in format and construction. Text annotation and link integration both use a textual expression to describe the business rules and connect them with the corresponding section of the process model. However, text annotation can result in repetition and, consequently, inconsistency of rule representation - i.e. the same rule being represented with slightly different text. For link integration, visual links can explicitly connect corresponding rules with the relevant process section. Even though link integration requires access to an external business rules repository, it is shown to reduce cognitive load required to mentally connect rules with process models [34]. Since the diagrammatic integration relies on graphical process model construction, such as, sequence flows and gateways, to represent business rules in the process model, limitations in representational capacity of the modeling language inevitably causes barriers or results in an increase in the complexity of the process model structures, which in turn may potentially result in an increase cognitive load for understanding the business process with rules integrated in diagrammatic format.

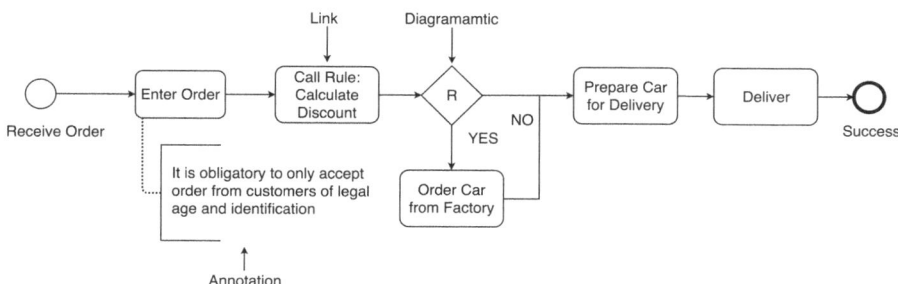

Fig. 1. Business rules integration approaches [34]

At the same time, a variety of factors have been identified as affecting the understanding of a process. These can be classified into two categories: process model factors and human factors. Process model factors relate to the metrics of the process models, such as modularization [28, 33], block structuredness [1, 38], and complexity. Human factors, or personal factors, relate to the factors of process model users, such as individual's domain knowledge [33], modeling knowledge [5], modeling experience [22], and education level [37].

A number of prior studies have focused on different forms of process model complexity, with a broad consensus that most complexities contribute to the decreased understandability of process models. The independent variables investigated in these works include: number of arcs and nodes [22, 26], number of gateways [28, 30, 31], number of events [30], number of loops [5], and number of concurrencies [21], length of the longest path [21, 31], depth of nesting [9], and gateway heterogeneity [21, 31]. For example, [19] studied the relationship between structural properties and process understandability. They mentioned that the number of arcs in models will influence the understandability, and later in [20], they presented a set of seven process modeling guidelines that can help modelers to create less error prone models. Similarly, [17] measured the understandability of process models, and among their findings for measurement in structural model comprehension, they argued that concurrency and exclusiveness are more complicated compared with order. Other researchers identified content related factors such as the separability, reliability and validity of model that can influence the process understandability [18, 21].

Another area of relevance for our work is cognitive load theory, which refers to the total amount of mental effort being used in working memory [23]. To perceive mental effort, researchers have categorized the measurement of cognitive load into four main aspects: subjective ratings, performance measures, behavioural measures and physio-logical measures [2]. Subjective measures, also referred to as self-report measures, use single or multiple rating scales used by the user to rank/score their experienced level of load; Performance measures consist of task completion time, answer correctness, etc.; Physiological measures involve tracking galvanic skin response and heart rate; and Behavioural measures involve observing patterns of interactive behaviour [2]. In practice, behavioural and physiological measures are often used as they provide a direct measurement of cognitive load. Among the various related measurements, eye-based measures are one of the main behavioural measurements as they can provide a sensitive and a reliable measure for cognitive load. Due to the limited working memory capacity and cognitive resources, we can conclude from prior research that a heavy cognitive load will lead to error in process model understanding, and that the error frequency will increase with the level of cognitive load [34]. Therefore, it is important to study the merits of integrating business rules into business process models in terms of its implications on cognitive load and subsequent improvement (or lack of) in the understanding of business process [34].

Eye tracking has emerged in recent years as one of the key sensor technologies applied in studies of visual cognition [4], and has enjoyed adoption by researchers across many fields. Based on the cognitive load theory, eye activity is one of the physiological variables that can be used as a technique to reflect the changes in cognition [4, 23]. Through the use of eye tracker technologies, such as the Tobii Pro TX300[1], we can directly collect eye movement data and measure objective metrics such as pupillary response and fixation durations to indicate the correlation with cognitive function [2]. By detecting indicators such as fixation in each area of interest (AOI), we can directly identify the exact area that draws the attention of the participant. Although there is a long history on the use of eye tracking technologies in medical and

---

[1] For more specifications of eye tracker, please visit https://www.tobiipro.com/product-listing/tobii-pro-tx300/.

psychological studies [12], the use of such technology in the business process modeling context is quite recent. To name a few, Petrusel and Mendling [24] defined the notion of Relevant Region and Scan-path to prove that Relevant Region is correlated to the answer during question comprehension. In [11], researchers used eye tracking method to measure and assess user satisfaction in process model understanding. In [25], the use of eye tracking technology enabled the researchers to identify the visual cues of coloring and layout that can improve performance in process model understanding.

3   Research Design

We use an experimental research design to undertake empirical evaluation of the three approaches to business process model and business rule integration. Our business process modeling language of choice is BPMN 2.0, due to its wide adoption and standing as an international process modeling standard. The experiment is inspired by methodologies proposed in [4], and has been adapted as explained below. Further, we consider the condition of lab environment, generalization ability and the need to control the learning effect during the experiment design.

The independent variable to be studied relates to the three approaches of business rules integration: text annotation, diagrammatic and link integration. The corresponding dependent variables are understanding accuracy, mental effort and time efficiency. Similar to other studies, we use measures of correctness of answers and time duration for answering questions to reflect the effectiveness of comprehension (or understanding accuracy) [6, 25]. Therefore, we use the number of correct answers to measure understanding accuracy. As for the time efficiency, the timing is counted starting from when the first process model is displayed on the computer screen, until the last question is answered and submitted. To measure participants' mental effort, we use fixation duration as the objective measure in this experiment, which is now increasingly used as a mental effort measure in lieu of pupil dilation [16].

The overall experiment design is illustrated in Fig. 2. Each group of participants is first provided a BPMN tutorial and is then offered two models but using one of the three different approaches of rule integration. One of our scenarios, on which the models and rules are based, originates from a travel booking diagram included in OMG's BPMN 2.0 examples[2]. The second model is adopted from Signavio website resources[3]. For the purposes of this study, we have ensured, through multiple revisions, that we have created informationally equivalent models for all three integration approaches. Due to space limitation, the models cannot be included in the paper, but the complete materials of entire experiment are available for download on Dropbox[4].

---

[2] Model originated from a travel booking diagram in OMG's BPMN 2.0 examples can be viewed in http://www.omg.org/cgi-bin/doc?dtc/10-06-02.

[3] Model adopted from Signavio website can be viewed in https://www.signavio.com/post/process-thinking-insurance/.

[4] The experiment materials can be download from https://www.dropbox.com/sh/eiow8c3z6u4vx7w/AACm44dstgRm2KRLJBRzwF8Na?dl=0.

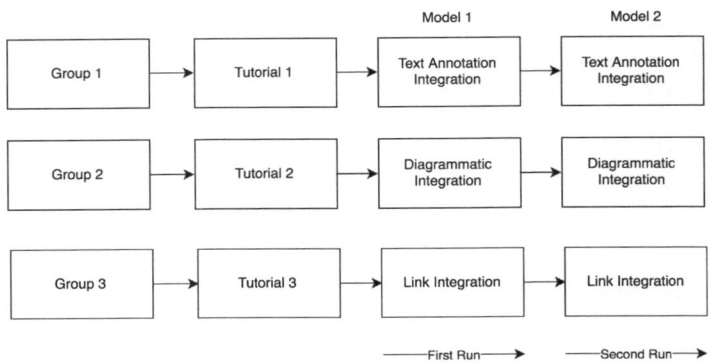

Fig. 2. Overall experiment design

In the remaining section, we introduce the instruments, settings and participants of our experiment.

3.1 Instruments

In this research, the instruments we use include a tutorial, the treatments and a questionnaire. In addition, we ensure all other confounding factors are constant, such as same eye-tracking lab equipment and same tutorial content. We do not impose a time limit or word count limit on the participants. In the treatment, we used the three integration approaches across two models. The two models are independent, and from different knowledge domains, however we made every attempt to maintain information equivalence and comparable complexity between them. Both models were adjusted to ensure consistency of format for each of the integration approaches. The two models have some diversity in terms of model constructs, for example the diagrammatic integration approach of model 1 has more parallel gateways (AND gateways, 18 vs. 6 in model 2), whereas model 2 has more exclusive gateways (XOR gateways, 15 vs. 3 in model 1).

In Table 1 we outline this diversity in terms of model constructs and model coverage for each question in each model. The listed model constructs indicate which constructs a participant will have to review in order to answer that question. Model coverage relates to the span of the question wherein a participant may have to navigate only a specific section of the process model to answer the question (local), or the whole process (global). We deliberately introduced diversity in questions to explore how each integration approach will affect cognitive load depending on process characteristics. This diversity allowed us to gain further insights into the relationship between process model constructs, rule integration approach and cognitive load (further details in the results section).

We designed the tutorial and tutorial exercises to help participants develop familiarity with BPMN and the format of the main experiment. Since this experiment does not require any substantial knowledge from participants, only basic BPMN constructs are used in the tutorial and experiment models. The tutorial was presented at the

Table 1. Comparison of questions

Model	Question	Model constructs	Model coverage
1	Q1	Sequence, AND gateways	Local area
	Q2	Sequence, AND gateways	Local area
	Q3	Sequence, AND gateways, XOR gateways	Global and local areas
2	Q1	Sequence, AND gateways, XOR gateways	Local area
	Q2	Sequence, AND gateways, XOR gateways, Loops	Local area
	Q3	Sequence, AND gateways, XOR gateways, Loops	Global and local areas

beginning of the experiment session and we encouraged each participant to ask any questions during the tutorial session, so as to ensure their readiness for the experiment.

To keep a group balance, we used a pre-experiment questionnaire to determine participants' prior knowledge and basic demographics to distribute participants to each group in a way that avoids accidental homogeneity of groups [4].

3.2 Setting

In this experiment, the questionnaire was implemented in Google Forms. The tutorial and experiment were carried out in an online web platform by using HTML, CSS, JavaScript and PHP with a back-end database using phpMyAdmin. The Areas of Interest (AOI) are created in Tobii Studio as shown in Fig. 3.

For the purpose of faithfully recording the eye tracking data, the experiment webpage was in full screen mode and complete models were displayed, without the function of zooming in or scrolling as these were not necessary. During the pilot test, the visibility of the experiment text and diagrams were examined carefully, and we ensured that all text and diagrams were clear from a distance of 1.2 m.

To eliminate colour blindness bias, we used a black, white and grey colour scheme for the Rule icon in link integration model. In addition, all experiments were conducted in the same lab with the same equipment. The lab is a small room with only a few machines and no windows, with a ceiling light as the only light source. The eye tracker equipment used in the experiment is the Tobii Pro TX300, with 23-inch screen of a resolution of 1920 × 1080. The participants were able to adjust the chair height to have the most comfortable position before calibration.

We used multiple Areas of Interest (AOI) to capture eye movements. For models featuring text annotation and diagrammatic integration, the screen was divided into two areas: a process model area and a question area. The process model area displayed the business process model, and the question area contained one question at a time for each model. For models featuring link integration, there was an additional third area for rules, which displayed the corresponding business rules when participants clicked on each "R" icon in the model, as shown in Fig. 3.

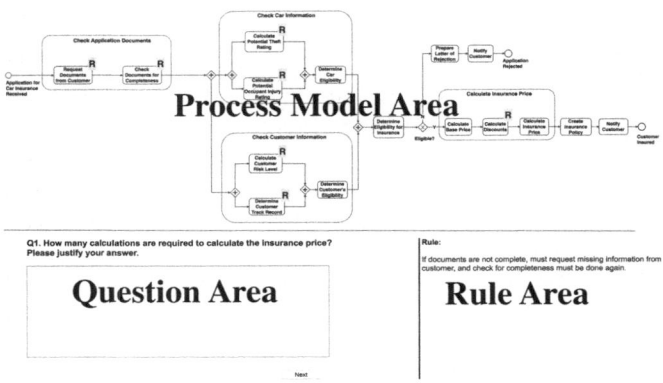

Fig. 3. Instrument illustrations of link integration

To ensure good quality of resulting data in the analysis of eye-movement related data, we had to eliminate the data of three participants whose eye movements failed to be properly recorded by Tobii eye tracker, that is, the eye tracker lost track of participant's eyes and the data did not faithfully reflect the fixation of eye movement.

3.3 Participants

All participants were students invited from an Australian university. They were required to have only foundational knowledge in graphical conceptual models such as flowcharts, UML or ER diagrams, but were not required to have any substantial knowledge of business process or rule modeling. Participation was on a voluntary basis, but participants were offered a $30 voucher for participating in this research. There were 25 participants in each group, with experiments conducted one at a time. In total, 75 students participated in this experiment. As in other similar experiments [10, 16], the sample size of 20 to 30 participants for each group is feasible, providing us with sufficient volume of data for testing statistical significance.

4   Results

Our data analysis is focused on understanding accuracy, time efficiency and mental effort. We use the number of correct answers of each participant (ordinal data) as a measure of understanding accuracy. For mental effort and time efficiency, we use fixation duration and visit duration (numerical data) based on eye tracking data. We structured our analysis into three different levels to draw out the subtle differences: overall results for the dependent variable, model level results, and question level results.

The approach taken for the analysis of the data we captured in the experiments is outlined as follows: For numerical data, we first use Shapiro-Wilk test[5] to check whether the dependent variable is normally distributed. If data is normally distributed, we use Levene's test[6] for homogeneity of variance to check whether it can meet the assumption of equal variance. If both the conditions are met, we use one-way analysis of variance (ANOVA) to further test the difference of means in the three groups. If there is a significant difference between the dependent variable and the integration groups, we use Tukey's HSD[7] as the post-hoc test to further compare the difference in each pair of groups. If normality is violated, we use the Kruskal-Wallis test[8]. If the Kruskal-Wallis test result is significant, we use the Dunn's test[9] to rank the groups in a pair-wise comparison, as it is a commonly used post-hoc test of Kruskal-Wallis test. The Bonferroni correction was not used because the independent variable has three groups. For ordinal data, we use Kruskal-Wallis test. If the result is significant, we use the same post hoc test to rank the groups in pair-wise comparison. The significance level of 0.05 is used in all the tests.

4.1   Understanding Accuracy

We first investigate whether there is a relationship between the rule integration approach and understanding accuracy, captured through correctness of answers.

Overall the result of Kruskal-Wallis test indicates that there is a significant difference between the three groups in terms of understanding accuracy ($p = 0.000$). The result of post-hoc pairwise comparisons show that diagrammatic integration is associated with higher understanding accuracy than text annotation (one-tailed $p = 0.000$) and link integration (one-tailed $p = 0.003$), but that text annotation and link integration do not differ significantly (one-tailed $p = 0.139$).

Model Level: As the results of Kruskal-Wallis test show in Table 2, we can conclude that there is a significant difference between the three groups in terms of understanding accuracy, both in model 1 ($p = 0.002$) and model 2 ($p = 0.033$). Given the result of post-hoc pairwise comparisons, we can further conclude that diagrammatic integration is associated with higher understanding accuracy than text annotation and link integration in both models, at the significance level of 0.05.

Question Level: From Fig. 4, we can observe that there is a notable contrast in the mean comparison in understanding accuracy between diagrammatic integration and the other two approaches in the first two questions in model 1 and the last two questions in model 2.

---

[5] The Shapiro-Wilk test is a test of normality.

[6] Levene's test is an inferential statistic used to assess the equality of variances for a variable calculated for two or more groups.

[7] Tukey's HSD is a post-hoc analysis of ANOVA that can be used to find means that are significantly different from each other.

[8] Kruskal-Wallis test is a non-parametric method when there are more than two groups.

[9] Dunn's test is a non-parametric multiple comparison post-hoc test of Kruskal-Wallis test.

Table 2. Understanding accuracy

Model	Group	N	Mean	Std. dev	p	Rank	p (1-tailed)
1	text	25	1.20	0.763	0.002	Diagrammatic > Text	**0.000**
	diagrammatic	25	2.08	0.862		Diagrammatic > Link	**0.009**
	link	25	1.48	0.918		Link > Text	0.145
2	text	25	0.72	0.891	0.033	Diagrammatic > Text	**0.005**
	diagrammatic	25	1.28	0.737		Diagrammatic > Link	**0.044**
	link	25	0.88	0.781		Link > Text	0.196

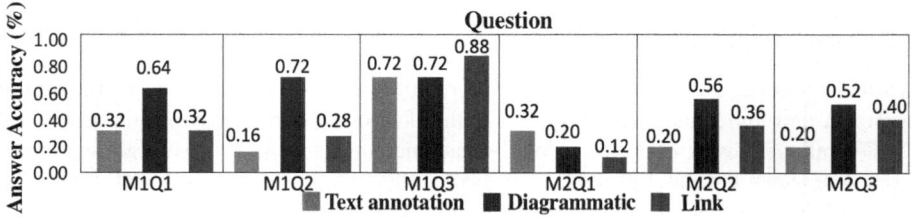

Fig. 4. Understanding accuracy breakdown to each question

Conclusion 1: Understanding accuracy is associated with the rule integration approach. Overall the diagrammatic integration shows better understanding accuracy than link and text integration. The same applies in model 1 and model 2. Link integration and text annotation do not significantly differ in understanding accuracy in all models.

4.2   Mental Effort

Overall, the result of Kruskal-Wallis test indicates that the difference in fixation duration between the three groups is not statistically significant ($p = 0.082$). Therefore, we further analyse the data relating to each model and each question to explore any detailed differences.

Model Level: The results of Shapiro-Wilk test for mental effort, indicate that the assumption of normality in both models are not met, both $p < 0.05$ ($p = 0.000$ and $p = 0.037$). Hence, we use Kruskal-Wallis test in both models. As shown in Table 3, the difference in fixation duration between the three groups in model 1 is not statistically significant ($p = 0.946$). As for the results of model 2, our analysis indicates that the difference in fixation duration across three groups is statistically significant ($p = 0.036$).

Table 3. Mental effort

Model	Group	N	Mean	Std. dev	p
1	Text annotation	24	227.550	100.998	0.946
	Diagrammatic	24	223.636	98.933	
	Link	24	221.671	97.808	
2	Text annotation	24	215.449	92.896	**0.036**
	Diagrammatic	24	266.484	90.554	
	Link	24	209.951	105.325	

For model 2, the result of post-hoc pairwise comparisons shows that the diagrammatic integration group has a statistically significant higher fixation duration than text annotation (one-tailed $p = 0.021$) and link integration (one-tailed $p = 0.008$), but text annotation and link integration do not differ significantly (one-tailed $p = 0.359$).

Question Level: From the mean comparison in fixation duration of each integration approach in Fig. 5, we can observe that there is a notable difference between diagrammatic integration and the other two integration approaches in the last two questions of model 2. We note that both these questions involved loop constructs.

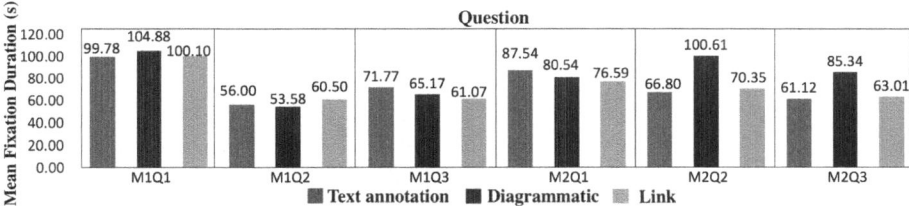

Fig. 5. Fixation duration breakdown to each question

Conclusion 2: Mental effort is partially associated with the rule integration approach. Overall, there is no significant difference in mental effort between different integration approaches. The same applies in model 1. In model 2 diagrammatic integration requires more mental effort than other integration approaches, especially when loop constructs are involved. Text annotation and link integration do not differ significantly in mental effort in all models.

4.3 Time Efficiency

Overall the result of Kruskal-Wallis test indicates that the difference in fixation duration between the three groups is not statistically significant ($p = 0.273$).

Model Level: The results of Shapiro-Wilk test for time efficiency, indicate that the assumption of normality in both models are not met, all with $p < 0.05$ ($p = 0.000$ and $p = 0.014$). Therefore, we use Kruskal-Wallis test in both models. As shown in

Table 4, for model 1 the result indicates that the difference in time efficiency between the three groups is not statistically significant ($p = 0.884$). In model 2, the difference in time efficiency across three groups is statistically significant ($p = 0.021$).

Table 4. Time efficiency

Model	Group	N	Mean	Std. dev	p
1	Text annotation	24	282.404	125.179	0.884
	Diagrammatic	24	281.908	124.974	
	Link	24	274.167	128.258	
2	Text annotation	24	264.210	114.704	**0.021**
	Diagrammatic	24	332.147	108.640	
	Link	24	257.391	127.388	

For model 2, the result of pairwise comparisons shows that the diagrammatic integration group has a statistically significant higher visit duration than text annotation (one-tailed $p = 0.012$) and link integration (one-tailed $p = 0.006$). However, text annotation and link integration group do not differ significantly (one-tailed $p = 0.394$).

Question Level: From the mean comparison of visit duration in each integration approach in Fig. 6, we can observe that there is a notable difference between diagrammatic integration and the other two integration approaches in the last two questions of model 2, which involved loop constructs.

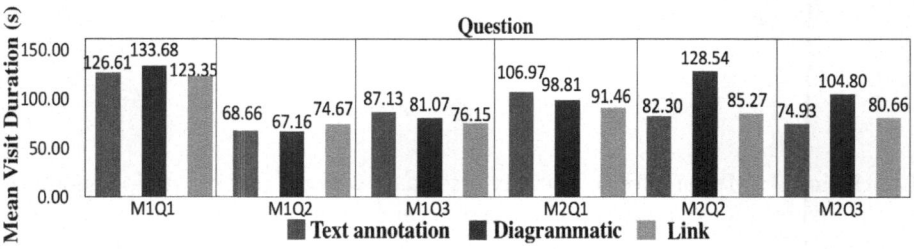

Fig. 6. Visit duration breakdown to each question

Conclusion 3: Time efficiency is partially associated with the rule integration approach. Overall there is no significant difference in time efficiency between the different integration approaches, nor in model 1 when considered in isolation. In model 2, diagrammatic integration requires more time than other integration approaches when loop constructs are involved in the questions. Text annotation and link integration do not differ significantly in time efficiency in all models.

4.4   Analysis and Discussion

Overall, we can observe that for model 1, diagrammatic integration is more under-standable than text annotation and link integration, but there is no significant difference in mental effort and time between the different integration approaches. At the same time, in model 2, diagrammatic integration is more understandable than text annotation and link integration, but requires more effort and time than the other two types of integrations. In reviewing these results against the diversity of model constructs and coverage (as outlined in Table 1), we stipulate that the differences in model constructs are the likely cause of these results. From a model construct perspective, we observe that model 2 has relatively more XOR gateways compared to model 1 in the dia-grammatic integration approach (15 vs. 3). In model 2, the last two questions are focused on the model area that involves looping. However, there is no looping in model 1 and the other questions do not require the participant to mentally navigate gateways. Hence, we note that the presence and number of XOR and AND gateways, and loops formed through these constructs, may influence the mental effort and time efficiency. Meanwhile, as shown in Figs. 5 and 6, there is a notable difference in the last two questions of model 2 in terms of diagrammatic integration requiring most mental effort and time. We posit that the increase in mental effort and time we observed in model 2 is attributed to the number of XOR and AND gateways, and loops formed through these constructs, as mentioned above. Moreover, we believe the reason that diagrammatic integration requires more mental effort and time than the other two approaches in model 2, is that it uses more gateways than link/text annotation integration (21 vs. 7) to integrate business rules into the model, which inevitably causes the model to become more complex.

Based on above analysis, we consider mental effort and time efficiency to be only partially associated with the rule integration approach. That is, diagrammatic integra-tion is associated with better understanding accuracy, but may require more mental effort and time than text annotation and link integration when the model involves complex loop constructs.

5   Conclusion

The central question in this research is to explore the difference between business process and rule integration approaches on business process model understanding. We set out to investigate this question through a cognitive load perspective by using eye-tracking, and studied the difference in terms of understanding accuracy, mental effort and time efficiency. Through the analysis, we discovered that the integrated approaches applied to models with specific characteristics will impact on cognitive load and consequently process understanding. For example, the presence and quantity of XOR gateways, AND gateways, and questions which require navigation of constructs through loop structures, seems to influence understanding, as observed in model 2.

The findings of this research provide empirical evidence of the relative merits of integration approaches. These findings can help process modelers make evidence-based decisions regarding integration of rules and process models relative to model characteristics. The design of this research experiment can also provide valuable methodological contribution in the field of business process model understanding. In particular, we illustrate feasible protocols and resulting advantages of using eye-tracking to study business process model understanding.

Our study is not without limitations. First, due to the limitation of the eye tracking software and the display capacity of the screen, the complexity of the process models and rules was restricted. Second, only two models were used in the experiment, which may hamper the generalizability of the conclusions. Further both models were created using BPMN, which also raises a question of generalizability across other business process modelling notations. Third, the validity of the results is potentially compromised by learning effects, since model 2 was assessed after model 1 in all experiments. Lastly, fatigue can also be considered as a potential weakness as there was no break for participants between the two models and we had no time limit for each participant to answer each question. Moreover, the individual variability (e.g. experience and domain knowledge) may influence the experiment results. Since all participants were students, we limit generalizability of the research to novice modelers. While organizational models are often more complex in reality, our findings still provide valuable comparative evaluations towards understanding the differences between integration approaches.

In our future work, we seek to extend our study with a consideration of further diversity of model construction and model coverage of rules, to better understand under what conditions the three different integration approaches perform better. We will design structural characteristics of the models in a way that enables us to measure effects of specific constructs on the dependent variables. We also plan to investigate the relationship between the dependent variables. Finally, this work can be extended to alternative process modeling notations, that is beyond BPMN, as different notations have different mechanisms for integrating rules, which is likely to effect process model understanding.

References

1. Burton-Jones, A., Meso, P.N.: Conceptualizing systems for understanding: an empirical test of decomposition principles in object-oriented analysis. Inf. Syst. Res. **17**, 38–60 (2006)
2. Chen, F., et al.: Robust Multimodal Cognitive Load Measurement. HIS. Springer, Cham (2016). https://doi.org/10.1007/978-3-319-31700-7
3. Cheng, R., Sadiq, S., Indulska, M.: Framework for business process and rule integration: a case of BPMN and SBVR. In: Abramowicz, W. (ed.) BIS 2011. LNBIP, vol. 87, pp. 13–24. Springer, Heidelberg (2011). https://doi.org/10.1007/978-3-642-21863-7_2
4. Duchowski, A.T.: Eye Tracking Methodology: Theory and Practice. Springer, London (2007). https://doi.org/10.1007/978-1-84628-609-4
5. Figl, K., Laue, R.: Influence factors for local comprehensibility of process models. Int. J. Hum. Comput. Stud. **82**, 96–110 (2015)
6. Gemino, A., Wand, Y.: A framework for empirical evaluation of conceptual modeling techniques. Requir. Eng. **9**(4), 248–260 (2004)

7. Governatori, G., Shek, S.: Rule based business process compliance. In: Proceedings of the RuleML2012@ ECAI Challenge, article 5 (2012)
8. Green, P.F., Rosemann, M.: Perceived ontological weaknesses of process modeling techniques: further evidence. In: Proceedings of the ECIS, pp. 312–321 (2002)
9. Gruhn, V., Laue, R.: Adopting the cognitive complexity measure for business process models. In: 5th IEEE International Conference on Cognitive Informatics (ICCI 2006), Beijing, China (2006)
10. Haji, F.A., Rojas, D., Childs, R., de Ribaupierre, S., Dubrowski, A.: Measuring cognitive load: performance, mental effort and simulation task complexity. Med. Educ. **49**, 815–827 (2015)
11. Hogrebe, F., Gehrke, N., Nüttgens, M.: Eye tracking experiments in business process modeling: agenda setting and proof of concept. In: Proceedings of EMISA 2011, pp. 183–188 (2011)
12. Just, M.A., Carpenter, P.A.: Eye fixations and cognitive processes. Cogn. Psychol. **8**, 441–480 (1976)
13. Kappel, G., Rausch-Schott, S., Retschitzegger, W.: Coordination in workflow management systems — a rule-based approach. In: Conen, W., Neumann, G. (eds.) ASIAN 1996. LNCS, vol. 1364, pp. 99–119. Springer, Heidelberg (1998). https://doi.org/10.1007/BFb0027102
14. Knolmayer, G., Endl, R., Pfahrer, M.: Modeling processes and workflows by business rules. In: van der Aalst, W., Desel, J., Oberweis, A. (eds.) Business Process Management. LNCS, vol. 1806, pp. 16–29. Springer, Heidelberg (2000). https://doi.org/10.1007/3-540-45594-9_2
15. McBrien, P., Seltveit, A.H.: Coupling process models and business rules. In: Sölvberg, A., Krogstie, J., Seltveit, A.H. (eds.) Information Systems Development for Decentralized Organizations. ITIFIP, pp. 201–217. Springer, Boston (1995). https://doi.org/10.1007/978-0-387-34871-1_12
16. Meghanathan, R.N., van Leeuwen, C., Nikolaev, A.R.: Fixation duration surpasses pupil size as a measure of memory load in free viewing. Front. Hum. Neurosci. **8**, 1063 (2015)
17. Melcher, J., Mendling, J., Reijers, H.A., Seese, D.: On measuring the understandability of process models. In: Rinderle-Ma, S., Sadiq, S., Leymann, F. (eds.) BPM 2009. LNBIP, vol. 43, pp. 465–476. Springer, Heidelberg (2010). https://doi.org/10.1007/978-3-642-12186-9_44
18. Melcher, J., Seese, D.: Towards validating prediction systems for process understandability: measuring process understandability (experimental results). Research report, Universität Karlsruhe (TH), Institut AIFB (2008)
19. Mendling, J., Reijers, H.A., Cardoso, J.: What makes process models understandable? In: Alonso, G., Dadam, P., Rosemann, M. (eds.) BPM 2007. LNCS, vol. 4714, pp. 48–63. Springer, Heidelberg (2007). https://doi.org/10.1007/978-3-540-75183-0_4
20. Mendling, J., Reijers, H.A., van der Aalst, W.M.P.: Seven process modeling guidelines (7PMG). Inf. Softw. Technol. **52**, 127–136 (2010)
21. Mendling, J., Strembeck, M.: Influence factors of understanding business process models. In: Abramowicz, W., Fensel, D. (eds.) BIS 2008. LNBIP, vol. 7, pp. 142–153. Springer, Heidelberg (2008). https://doi.org/10.1007/978-3-540-79396-0_13
22. Mendling, J., Strembeck, M., Recker, J.: Factors of process model comprehension—findings from a series of experiments. Decis. Support Syst. **53**, 195–206 (2012)
23. Paas, F., Tuovinen, J.E., Tabbers, H., Van Gerven, P.W.M.: Cognitive load measurement as a means to advance cognitive load theory. Educ. Psychol. **38**, 63–71 (2003)
24. Petrusel, R., Mendling, J.: Eye-tracking the factors of process model comprehension tasks. In: Salinesi, C., Norrie, M.C., Pastor, Ó. (eds.) CAiSE 2013. LNCS, vol. 7908, pp. 224–239. Springer, Heidelberg (2013). https://doi.org/10.1007/978-3-642-38709-8_15
25. Petrusel, R., Mendling, J., Reijers, H.A.: Task-specific visual cues for improving process model understanding. Inf. Softw. Technol. **79**, 63–78 (2016)

26. Recker, J.: Empirical investigation of the usefulness of Gateway constructs in process models. Eur. J. Inf. Syst. **22**, 673–689 (2012)
27. Recker, J., Rosemann, M., Green, P.F., Indulska, M.: Do ontological deficiencies in modeling grammars matter? MIS Q. **35**, 57–79 (2011)
28. Reijers, H., Mendling, J., Dijkman, R.: Human and automatic modularizations of process models to enhance their comprehension. Inf. Syst. **36**, 881–897 (2011)
29. Rima, A., Vasilecas, O., Šmaižys, A.: Comparative analysis of business rules and business process modeling languages. Comput. Sci. Tech. **1**(1), 52–60 (2013)
30. Rolón, E., Garcia, F., Ruiz, F., Piattini, M., Visaggio, C.A., Canfora, G.: Evaluation of BPMN models quality - a family of experiments. In: ENASE, pp. 56–63 (2008)
31. Sánchez-González, L., García, F., Mendling, J., Ruiz, F.: Quality assessment of business process models based on thresholds. In: Meersman, R., Dillon, T., Herrero, P. (eds.) OTM 2010. LNCS, vol. 6426, pp. 78–95. Springer, Heidelberg (2010). https://doi.org/10.1007/978-3-642-16934-2_9
32. Sapkota, B., van Sinderen, M.: Exploiting rules and processes for increasing flexibility in service composition. In: 2010 14th IEEE International Enterprise Distributed Object Computing Conference Workshops (EDOCW), pp. 177–185. IEEE (2010)
33. Turetken, O., Rompen, T., Vanderfeesten, I., Dikici, A., van Moll, J.: The effect of modularity representation and presentation medium on the understandability of business process models in BPMN. In: La Rosa, M., Loos, P., Pastor, O. (eds.) BPM 2016. LNCS, vol. 9850, pp. 289–307. Springer, Cham (2016). https://doi.org/10.1007/978-3-319-45348-4_17
34. Wang, W., Indulska, M., Sadiq, S.: Cognitive efforts in using integrated models of business processes and rules - semantic scholar. In: Proceedings of the 28th International Conference on Advanced Information Systems Engineering (CAiSE Workshop), Ljubljana, Slovenia. Springer (2016)
35. Wang, W., Indulska, M., Sadiq, S.: To integrate or not to integrate – the business rules question. In: Nurcan, S., Soffer, P., Bajec, M., Eder, J. (eds.) CAiSE 2016. LNCS, vol. 9694, pp. 51–66. Springer, Cham (2016). https://doi.org/10.1007/978-3-319-39696-5_4
36. Wang, W., Indulska, M., Sadiq, S., Weber, B.: Effect of linked rules on business process model understanding. In: Carmona, J., Engels, G., Kumar, A. (eds.) BPM 2017. LNCS, vol. 10445, pp. 200–215. Springer, Cham (2017). https://doi.org/10.1007/978-3-319-65000-5_12
37. Weitlaner, D., Guettinger, A., Kohlbacher, M.: Intuitive comprehensibility of process models. In: Fischer, H., Schneeberger, J. (eds.) S-BPM ONE 2013. CCIS, vol. 360, pp. 52–71. Springer, Heidelberg (2013). https://doi.org/10.1007/978-3-642-36754-0_4
38. Zugal, S., Pinggera, J., Weber, B., Mendling, J., Reijers, H.A.: Assessing the impact of hierarchy on model understandability – a cognitive perspective. In: Kienzle, J. (ed.) MODELS 2011. LNCS, vol. 7167, pp. 123–133. Springer, Heidelberg (2012). https://doi.org/10.1007/978-3-642-29645-1_14
39. Zur Muehlen, M., Indulska, M., Kittel, K.: Towards integrated modeling of business processes and business rules. In: Proceedings of the 19th Australasian Conference on Information Systems (ACIS)-Creating the Future: Transforming Research into Practice, Christchurch, New Zealand, pp. 690–697. Citeseer (2008)

Business Process Activity Relationships: Is There Anything Beyond Arrows?

Greta Adamo[1,4(✉)], Stefano Borgo[2], Chiara Di Francescomarino[1],
Chiara Ghidini[1], Nicola Guarino[2], and Emilio M. Sanfilippo[3]

[1] FBK-IRST, Trento, Italy
{adamo,dfmchiara,ghidini}@fbk.eu
[2] ISTC-CNR Laboratory for Applied Ontology, Trento, Italy
{stefano.borgo,nicola.guarino}@cnr.it
[3] Ecole Centrale de Nantes – Laboratoire des Sciences du Numérique – LS2N,
Nantes, France
emilio.sanfilippo@ls2n.fr
[4] University of Genova, DIBRIS, Genova, Italy

Abstract. Business process modelling languages enable the depiction of the processes of an organisation by exploiting graphical symbols to denote the key elements to be represented. Despite the variety of approaches, graphical symbols, and (in)formal interpretations associated to the different languages, a fundamental component of every business process modelling language is the representation of the way activities are related by means of control arcs and gateways. While *multiple* kinds of relationships may hold among such activities, mainstream business process modelling languages seem actually only interested in modelling a *single* (very important) kind of relationship, namely the *activity execution order* within the control flow. In this paper we investigate the role of another kind of fundamental relationship between activities, namely *ontological dependence*, in the context of business process modelling. In particular, we introduce three forms of generic ontological dependence, namely *historical* dependence, *causal* dependence, and *goal-based co-occurrence*. We illustrate different forms in which they can occur, we introduce a language to express them and we discuss their usefulness in two concrete use cases.

1 Introduction and Motivations

Business process modelling languages enable the depiction of the processes of an organisation by exploiting graphical symbols to denote the key elements to be represented. Examples are the sequence of activities to be executed (the so-called control flow), the actors involved, the data objects required/manipulated by the activities, message exchanges, and so on.

Despite the variety of approaches, graphical symbols, and their (in)formal interpretations, a fundamental component of every business process modelling language is the representation of the way activities (and events) are related by

© Springer Nature Switzerland AG 2018
M. Weske et al. (Eds.): BPM Forum 2018, LNBIP 329, pp. 53–70, 2018.
https://doi.org/10.1007/978-3-319-98651-7_4

means of control arcs and connectors (gateways). However, while mainstream business process modelling languages seem actually only interested in modelling a *single* (very important) kind of relationship, namely the *activity execution order* within the control flow, *multiple* kinds of other relationships may hold among such activities.

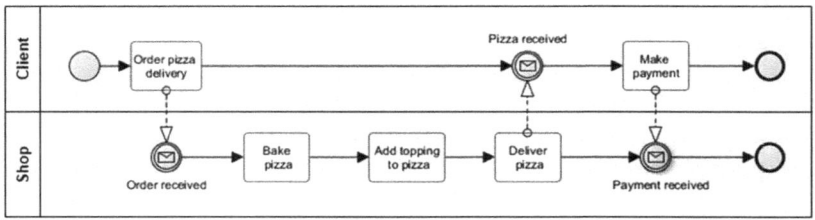

Fig. 1. A simple pizza delivery process model.

Consider, for instance, the simple BPMN diagram of Fig. 1. Its control arcs specify that the execution of a pizza delivery process starts with the order, continues with the baking of the pizza, the addition of toppings, the delivery, and the payment. In addition to the relation between activities captured by the control arcs, most human beings would easily identify further relationships in this process. As a first example, the (indirect) relationship between `Bake pizza` and `Deliver pizza` presupposes an intrinsic execution order that is independent on this particular process model. Indeed, delivering a pizza requires having (made) it first. This relation does not depend upon the way the organisation decides to structure the control flow. On the contrary, it holds in virtue of the very nature of such activities in the *real world*, and this influences the way the real business processes are organised (and thus represented in the model).

As a second example, one may notice that `Deliver pizza` and `Make payment` exhibit a different kind of mutual relationship. Indeed, an organisation may freely organise its own processes asking for payment before or after a delivery. We can nonetheless assume that the commercial nature of the pizza shop and its business goal of making money suggests that delivering a pizza must be (sooner or later) associated to a payment in order to have a meaningful process. These two simple examples show different real world relations between activities that can hold in the real world. Nonetheless, they are represented in the same way in the process model of Fig. 1. This happens because the model only represents the execution order of activities within the control flow.

The inability to account for aspects coming from *real world* constraints makes the standard business process modelling notations less informative from an explanatory perspective, and less robust against possible changes that violate fundamental domain constraints. Indeed, while the (intentionally very simple) pizza shop example reflects characteristics of the real world that most of us know, intrinsic aspects of more complex domains may be more difficult to understand,

and could be missed by people (or algorithms) who lack the background knowledge required to understand them.

This may result in crucial modelling mistakes, especially during model redesign. For example, while common sense would prevent refactoring the pizza process by imposing to deliver a pizza before baking it, no information in the model actually forbids that. Similarly, removing the payment activity from the process would dramatically change its meaning, so as to question whether it should still be considered the "same" process. On the contrary, the removal of the Add topping to pizza activity would intuitively be considered just a process refactoring.

Characterising relationships between business process activities beyond the control flow perspective is not trivial, due to the multitude of aspects and features that may be considered. For example, activity relationships may be distinguished according to their temporal features, co-occurrence constraints, the nature of actors or participants involved, and the goals of the business process. Some of these aspects reflect *normative choices* (or *business rules*) concerning the expected process structure, while others are bound to genuine *ontological constraints* intrinsic in the activities themselves. In the remaining of the paper we provide an analysis of some of such constraints, focusing in particular on temporal co-occurrence, and we apply them to distinguish among different kinds of activity relationships within business process models. In particular, we provide:

- an analysis of activity co-occurrences in terms of ontological dependences which allows to select and introduce three forms of generic ontological dependence among activities, namely *historical* dependence, *causal* dependence, and *goal-based co-occurrence* (Sect. 2);
- a first investigation of different forms of historical dependence, causal dependence, and goal-based co-occurrence, depending on their genuine *ontological* aspects, *goal*-related aspects, and *norm*-related aspects (Sect. 3);
- a proposal on how to incorporate historical dependence, causal dependence and goal-based co-occurrence in business process models by following a hybrid modelling approach (Sect. 4);
- an illustration of the usefulness of historical dependence, causal dependence, and goal-based co-occurrence in two concrete use cases concerning business process documentation and business process redesign (Sect. 5).

2   Activity Co-occurrence as Ontological Dependence

The goal of this paper is to make explicit the nature of the links holding amongst activities that pertain a business process. We rely here on Weske's definition [30], according to which a business process is *"a set of activities that are performed in coordination in an organizational and technical environment. These activities jointly realize a business goal. Each business process is enacted by a single organization, but it may interact with business processes performed by other organizations."*

In particular we based our analysis on *ontological dependences* resulting in co-occurrence constraints involving activities that occur during the same process execution. Such constraints hold by necessity in a particular domain, independently of the way a business process is designed. For example, delivering a pizza necessarily presupposes that the pizza has been baked. Similarly, no receive event can occur without a corresponding send event.

In formal ontology, ontological dependence is a fundamental relationship (or set of relationships) which can take many forms [5,10]. In general, an entity is dependent upon another when it is not ontologically *self-sufficient*, in the sense that it cannot exist alone. A basic form of dependence is so-called *specific* (or *rigid*) *existential dependence*, which holds among two objects when the existence of one necessarily implies the existence of the other. For instance, we may say that a person is specifically existentially dependent on her brain. A weaker form is the so-called *generic existential dependence*, which holds when the existence of an object requires the existence of another *of a given kind*. For instance, a human being is generically dependent on a heart (under the assumption that the heart may be substituted). An even weaker form of dependence may hold between kinds, when the existence of an instance of one kind requires the existence of an instance of the other kind. This seems to be enough in our case, since in most business process models key elements (such as activities in a BPMN model or transitions in a Petri Net) are indeed understood as *kinds*, and we are interested in the relationships among them. However, since the instances of such kinds are temporal entities, we should speak of *occurrence* instead of *existence*, so that instead of *existential dependence relationships* we have to talk of *co-occurrence dependence relationships*. In the following, we shall introduce three forms of ontological relations that characterize the nature of such co-occurrence dependence relationships. The reason why we have chosen these specific forms of ontological dependences between activities is twofold: on the one hand they are grounded on important generic ontological dependences investigated in literature; on the other hand they seem to play a fundamental role in all the business processes (models) that have been examined for this work.

A first type of co-occurrence dependence relationship is *historical dependence*. This captures the situation where a certain activity occurrence presupposes that another activity occurred *in the past*. For example, an instance of `Deliver pizza` may occur only if an instance of `Bake pizza` occurred beforehand. We shall define historical dependence as follows:

> Let P_1 and P_2 be business process activities (that is, *kinds* of actions that may occur in a business process). We shall say that P_1 is *historically dependent* on P_2 iff, necessarily, whenever an instance x of P_1 occurs at time t, there exists an instance y of P_2 that has occurred at a time $t' < t$.

Note that historical dependence is a relation holding *necessarily*, and has therefore an *ontological* nature. On the contrary, a mere *temporal precedence* relation simply resulting from the fact that two activities precede one another in a *particular* business process model may have just a *prescriptive* nature, if no historical dependence holds among the same activities. For example, a certain model may say that an activity Check contract should always precede the activity Sign contract. Although these activities may be done in any order (since none of them causes or implies the existence of the other), there is a clear reason to have them in a specific temporal order, but this reason reflects a *business rule* and not an ontological constraint.

A stronger type of occurrence dependence relationship is *causal dependence*. Causality is notoriously challenging to define [11], and its complete characterisation is behind the purposes of this work. For our purposes, we assume the following definition, which characterizes causality in terms of *contribution to explanation*:

> A process activity P_1 is *causally dependent* on P_2 iff, necessarily, whenever an instance x of P_1 occurs, there exist an instance y of P_2 that occurs before x, whose occurrence *contributes to explain why x occurred*.

This definition is admittedly naive, but it seems to be enough for practical cases. For example, an event of message receiving occurs because an event of message delivering occurred. Analogously, a pizza delivering activity occurs because an ordering event occurred in the past, and not because a particular pizza was baked. So, the relation between Deliver pizza and Bake pizza is a historical dependence, while that between Deliver pizza and Order pizza delivery is a causal dependence. Of course, a causal dependence implies a historical dependence.

Finally, a third kind of occurrence dependence relationship is what we shall call *goal-based co-occurrence*[1]:

> Let G be a goal, typically associated to a certain business process. The process activities P_1 and P_2 are *goal-based co-occurrence* iff the occurrence of both P_1 and P_2 is necessary for the satisfaction of G.

Consider that no temporal constraint is imposed on P_1 and P_2, which may occur in whatever order. In other terms, we only say that, for the satisfaction of G, instances of P_1 cannot occur if instances of P_2 do not occur, and vice versa. Consider, for example, the activity Deliver pizza in Fig. 1. Given the nature of our process' goal, which may be stated as "Selling pizza", both Deliver pizza and Make payment (for the pizza) are necessary for the satisfaction of such goal, and they are therefore co-occurrent with respect to such goal. Assuming that no historical dependence holds necessarily between the two activities, a process re-factoring is possible, where the delivery occurs before the payment. What is necessary, however, is that the payment occurs sooner or later. Note that goal-based co-occurrence is symmetric, differently from the previous two relations.

---

[1] While co-occurrences may, in principle, be based on different elements, goals seem to play a fundamental role in co-occurrences in all the business processes (models) we have examined for this work. We leave the investigation of other forms of co-occurrences for future work.

Note that, for the sake of simplicity, we are considering here only relationships between pairs of activities. Nonetheless the dependences introduced in this section could be generalised to multiple activities or to process patterns / subprocesses.

3 Forms of Occurrence Dependence

As already stated in Sect. 1, dependence relationships between business process activities can be motivated by different aspects of the world a real process is embedded in. In this section we exemplify, by means of examples, the role that (i) genuine *ontological* constraints (hereafter 'laws of nature'), (ii) the *goal* of the process, and (iii) *norms* can play in determining historical dependence, causal dependence and goal-based co-occurrence. While the categories considered here are not meant to be exhaustive, they are of fundamental importance for the representation of business processes. Genuine ontological dependence exists because of the way the real world is structured and cannot be circumvented by business processes. Dependences related to the goal often refer, in our opinion, to the very nature of the process. They may be circumvented, but their violations may have dramatical effects on the meaningfulness of the process. Finally, laws and regulations often define a social world as important as the physical one for business processes. Also in this case, dependences may be violated but their violations have strong effects on the compliance of the process w.r.t. the normative world that regulates them (see e.g., [13]).

3.1 Historical Dependence

Historical dependence seems to play an important role in business process models and may come in different forms. A first example is provided by pairs of activities that pertain the "switch" between two complementary states such as turning on and off, entering and exiting and so on. A paradigmatic example in business process models is constituted by the activities Login and Logout from a web page in a session. While it is possible the login occurs without a logout, the opposite can not occur. If a logout does occur, then the login must have occurred. This is a particular case of historical dependence and is due to a 'law of nature' that can be generalised, as we said, to all changes between complementary and mutually exclusive binary states. Different examples still due to 'laws of nature' are the ones of Bake pizza and Deliver pizza discussed in previous sections, or the one of an administrative procedure of applying for a PhD position in which an applicant submits the PhD request (application form) to the PhD office, which is then checked for compliance to the submission rules. Submit PhD application and Check PhD application are connected together by a historical dependence as the PhD office can not check something that has not been submitted. By generalisation, the two forms of historical dependence mentioned here depend upon a 'law of nature' that determines that one can perform an activity on an artefact only if this artefact exists and is available.

An example of historical dependence related to the goal of the process is the one involving two `Make diagnosis` and `Propose treatment` activities in a healthcare process. While a diagnosis is not a genuine ontological constraint for the proposal of a treatment, the goal of the process of providing an effective (if not the best) cure to a patient triggers this historical dependence in a meaningful process.

A further example of historical dependence may be due to normative laws. For instance, in an on-line shopping purchase a `Login` activity may be a normative necessary pre-requisite for the execution of a `Purchase goods` activity, in order to certify the identity of the customer. Similarly to the example above, while a login is not ontologically needed for a customer in order to buy something, the social world determined by the norm imposes that a customer identification via `Login` is strictly necessary in order to accomplish an e-buy activity.

3.2 Causal Dependence

A first form of causal dependence, due to a sort of 'law of nature', is the one that holds between `Send` and `Receive` activities (events, in certain notations). Indeed the activity `Send message` not only is an existential requirement for `Receive message` to exist but it also causes the receipt of the message itself.

Further examples of causal dependence can be found if we focus on the goal of a business process. Consider again the pizza example. In this example `Order pizza delivery` causes several further activities in the process, and in particular `Deliver pizza`. Note that this is not due to a 'law of nature' but to the goal of the pizza shop, which is the one of making money by selling pizzas to customers and fulfilling their (customers) expectations. While causal dependence is also historical dependence the opposite does not hold as `Bake pizza` does not cause its delivery. Indeed a pizza (or any good) is not sold just because it is made but because someone asked for it.

Normative regulations can also refer to activities that are involved in a causal dependence. Consider for instance the activity `First use of software` and `Evaluate terms and conditions`. In this example, the first usage of a just installed software triggers the evaluation of terms and conditions and also motivates/explains why this activity occurs in a software installation process. Similarly to the above this is not due to a 'law of nature' but to normative requirements regulating the usage of artefacts (the software, in our case).

3.3 Goal-Based Co-occurrence

When it comes to the goal of the process, a typical example of goal-based co-occurrence is the one involving the activities `Deliver good` and `Pay for good` in the context of an economically motivated selling-oriented business process, of which `Deliver pizza` and `Make payment` (for pizza) in Fig. 1 is a specific example already illustrated in Sect. 4. As a further example, consider the annual evaluation process of an employer in a given organisation. Whenever the goal is to ensure a transparent and fair evaluation, a goal-based co-occurrence may involve two activities Send evaluation to Human Resources and Send evaluation to employer executed by the employer's boss. Indeed the provision of the evaluation to Human Resources is required to make the evaluation adopted by the organisation, while with the provision of the evaluation to the employer provides a possibility to highlight unfair treatments, and they are jointly required to achieve the overall goal.

4   Modelling Dependence Relationships in Business Processes

In Sects. 2 and 3 we have introduced the historical dependence, the causal dependence and the goal-based co-occurrence, and illustrated, by means of examples, their occurrence in typical business process scenarios. Here we introduce a simple language for expressing these dependences, investigate their meaning in terms of temporal properties, and make a proposal on how to include them in (hybrid) business process models.

First of all we define the syntax of dependence expressions. Let $T = \{T_1, \ldots, T_n\}$ be an alphabet of business process activities. A *dependence expression* is an expression of the form $\text{COOC}(T_i, T_j)$, $\text{HIST}(T_i, T_j)$, and $\text{CAUSE}(T_i, T_j)$, where $T_i, T_j \in T, i \neq j$.[2] Next, we need to understand what is the meaning of these expressions and what does it mean to enforce them upon a business process model.

A first question we need to clarify is whether dependence expressions concern a business process diagram (only) or execution paths. From the description of dependences provided in the previous sections, it is clear that they refer to process execution paths. Indeed when we state, e.g., that activities Deliver pizza and Make payment (for pizza) co-occur in a process model we do not simply intend that they both should appear in a diagram in whatsoever position of the control flow (perhaps as mutually exclusive choices) or none should, but also the more stringent constraint that each actual pizza production process execution must contain both or none. A similar reading holds for a historical or a causal dependence.

Since dependence expressions have effects on finite execution traces, a way to characterise (some of) their effects on process executions is to describe them using Linear-time Temporal Logic (LTL_f) with $f$inite execution semantics [6]. $\text{COOC}(T_i, T_j)$ states that either T_i and T_j co-occur in a process execution or they both do not appear. This corresponds, in LTL_f to the formula $\Diamond T_i \leftrightarrow \Diamond T_j$. $\text{HIST}(T_i, T_j)$ states that the execution of T_j necessarily requires a previous execution of T_i. An occurrence of T_i, nonetheless does not depend upon T_j. In particular, when T_j is not present in the trace, T_i can either occur or not. This corresponds, in LTL_f to the formula $\neg T_j \, \mathcal{W} \, T_i$. $\text{CAUSE}(T_i, T_j)$ states that the execution of T_j necessarily requires a previous execution of T_i and the previous

---

[2] We follow previous work in the area of BPM and focus on process models with no repeating activities, in the spirit of [1]. The investigation of dependences between repeated activities occurring in loops is left for future work.

execution of T_i is necessary to explain the execution of T_j. Thus both T_i and T_j must occur in the execution in this order (or none of them does). This corresponds, in LTL$_f$ to the formula $\neg T_j \, \mathcal{W} \, T_i \wedge \Box(T_i \rightarrow \Diamond T_j)$. Given this interpretation of dependence expressions, we can note that a causal dependence enforces also a goal-based co-occurrence and a historical dependence.

Note that the characterisation of dependence expressions provided above only concerns some necessary temporal properties that these expressions should enforce upon a process execution. A formal characterisation of historical dependence, causal dependence and goal-based co-occurrence, that takes into account also their ontological nature is left for future work.

Incorporating Dependence Expressions in (Hybrid) Process Models. Dependence expressions are not meant to be used on their own. Instead, they are thought of as expressions that complement a business process model and provide the ability to capture aspects from the real world (including the social world and goal oriented aspects) that otherwise would be lost. In particular, in case of procedural process models, such as BPMN models or WF-nets, we envisage a model of a real process P as composed of two separate (but related) parts: a procedural model (diagram) and a set of dependence expressions. This proposal is in line with several recent work in the BPM field (see e.g., [7, 18]) where so-called *hybrid models* are introduced as a way to combine a procedural component that describes all the allowed control flows in an imperative manner and a declarative component that describes only what should not be violated. The two parts are kept separated so as not to hamper the perceptual discriminability of the various model elements [20].

Given the characterisation of $\text{COOC}(T_i, T_j)$, $\text{HIST}(T_i, T_j)$, and $\text{CAUSE}(T_i, T_j)$ in terms of LTL$_f$ one may consider the idea of exploiting the declarative language DECLARE [23] to represent dependence expressions. Indeed, it is easy to note that the interpretation of the three expressions provided here creates a correspondence between $\text{COOC}(T_i, T_j)$, $\text{HIST}(T_i, T_j)$, and $\text{CAUSE}(T_i, T_j)$ and the DECLARE patterns *co-existence*(T_i, T_j), *precedence*(T_i, T_j), and *succession*(T_i, T_j), respectively (see Table 1, where the graphical notation and the formalisation in terms of LTL$_f$ of relevant DECLARE patterns is proposed). The exploitation of DECLARE would leverage an existing modelling language, thus avoiding the burden of a new notation. Moreover, the investigation proposed here could be seen as a sort of ontological grounding of specific DECLARE patterns. Nonetheless, we prefer not to commit to this proposal in this paper. In fact, flattening e.g., a causal relation onto a succession pattern would have three undesirable consequences: first, it would overload the meaning of DECLARE patterns with notions that are outside DECLARE (the notion of causality in this case); second, it would reduce ontological dependence to mere temporal patterns; third, it would 'transfer' to ontological dependences entailments that are only valid for temporal patterns. As an example, while *co-existence*(T_i, T_j) and *precedence*(T_i, T_j) entail *succession*(T_i, T_j), it would be incorrect to state that a goal-based co-occurrence and a historical dependence between two activities also force the validity of a causal dependence among them.

Table 1. Graphical notation and LTL formalisation of some declare templates.

TEMPLATE	FORMALIZATION	NOTATION	DESCRIPTION
Response(A,B)	$\square(A \rightarrow \lozenge B)$	A ●——▶ B	If A occurs, B must eventually follow
Precedence(A,B)	$\neg B \, \mathcal{W} \, A$	A ——▶● B	B can occur only if A has occurred before
Co-existence(A,B)	$\lozenge A \leftrightarrow \lozenge B$	A ●——● B	If B occurs, then A occurs, and viceversa
Succession(A,B)	response (A,B) $\wedge$ precedence (A,B)	A ●——▶● B	A occurs if and only if it is followed by B

Nevertheless, the formalisation of dependence expressions in terms of LTL$_f$ enables us to leverage existing techniques and tools (e.g., [7,16]) for the automated check and repair of a procedural model with respect to dependence expressions, at least for what concerns their temporal characterisation.

5   Application Scenarios

In this section we describe two application scenarios which could benefit of the analysis carried out in the previous sections: business process documentation and business process redesign.

5.1   Business Process Documentation

Business process models are often used by organizations as a means for documenting the procedures carried out. However, the information contained in the model sometimes is not enough in order to make clear the reasons why some parts of the process model have been designed in a certain way.

Let us consider a realistic scenario of an *Intake process for elderly patients with mental problems*, inspired by the procedure reported in [9] that describes the process carried out in a healthcare institution of the Eindhoven region. The *Intake* process starts when the institute receives a notice by the family doctor of the person who needs the treatment. The notice is answered, recorded and printed. The patient's folder is retrieved, if it already exists, or it is created, if the patient has never been registered in the healthcare information system, and the notice added to the patient's folder. Two intakers (a social-medical worker and a physician) are then assigned to the patient and the assignments stored in the system. Two cards containing information about the patient, one per intaker, are printed and handed out. Meanwhile, if needed, the medical file of the patient is requested to the patient's doctor and, whenever it is received, the document is added to the patient's folder. Once the medical file is available for the appointment, the patient can meet the intakers and is asked to pay the ticket. At the end of each of the two meetings, the patient's folder is enriched with the new information acquired by the intakers. When the documentation by each of the two intakers has been collected, it is evaluated and a treatment for the patient decided.

Figure 2 reports the *Intake* process described in BPMN and annotated with some hypothetical activity cycle time (including both processing and waiting time) as well as with the probability distribution of the alternative branches.

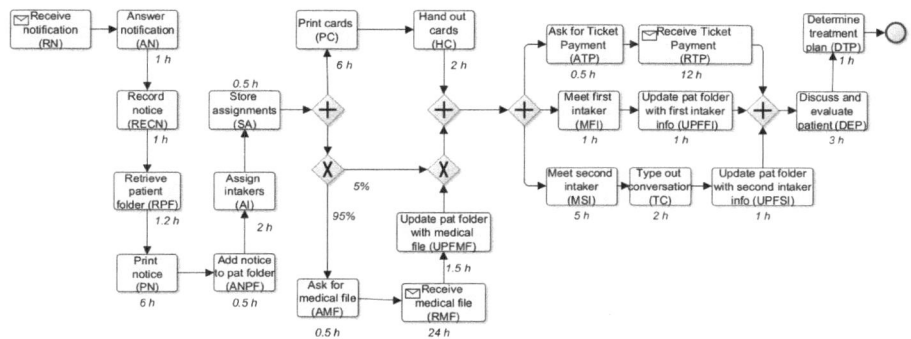

Fig. 2. *Intake* process of a healthcare institute

Let us assume that a new director has been appointed, and she has been provided with the institute business process models in order to get familiar with the procedures carried out in the institute. When looking at the *Intake* process model in Fig. 2 (in which data objects are not reported to ease the readability, and activity labels, as often happens, are not extremely informative), she is only able to grasp the execution ordering of the activities currently carried on in the institute, while missing other types of dependences among them. This lack of information could result in possible misunderstandings of the process model as well as of what it represents. By only looking at the model, she may ask the reason why in the model the activity `Assign intakers` occurs before the activity `Update pat (ient) file with first intaker information`.

Table 2 reports the dependence expressions identified among the activities of the *Intake* process. Some of the dependences are real-world ones, i.e., they depend on *laws of nature*, others relate to the *business goal* of the process, while others pertain to *norms*. The dependence expressions are grouped accordingly in Table 2.

Among the *law-of-nature* dependences, a historical dependence can be identified between the activities `Record notice` and `Print notice`. Intuitively, printing a notice demands for a state of the world in which the notice is in an electronic format, i.e., it requires that it has been (electronically) recorded. Similarly, a historical dependence exists between the `Retrieve patient folder` and all the activities that demand for the existence of the folder in order to be executed (i.e., `Add notice to patient folder`, `Update pat. folder with medical file`, `Update pat. folder with first intaker info`, `Update pat. folder with second intaker info`). A historical dependence also exists between the activities `Print cards` and `Hand out cards`, as handing out card demands for a state of the world in which the cards have been printed out. Few causal dependences can also be identified, as for instance between the activities `Receive notice` and `Answer notice` (the notice answer is caused or explained by the notice receipt), between the activities `Ask for medical file` and `Receive medical file` (the receipt of the medical file is caused by the request of the file to the doctor) and between the activities `Ask for ticket payment` and `Receive ticket payment` (the payment reception is caused by the payment request).

Among the *business goal* dependences, a goal-based co-occurrence can be identified between the activities `Receive ticket payment` and `Determine treatment plan`. Indeed, due to the business nature of the *Intake* process, in order to get the process accomplished, both determining the treatment plan for the patient and getting the ticket paid for the service are necessary activities. Removing the occurrence of one of the two activities would change the process into a different one. However, the two activities are not bound by any temporal constraint. Similarly, for the goal-based co-occurrence between the activities `Receive ticket payment` and `Discuss and evaluate patient info`. Moreover, a historical dependence can be identified between the activities `Discuss and evaluate patient` and `Determine treatment plan`. Indeed, in an *Intake* process, a decision on the treatment plan of a patient cannot be taken, unless the patient's information has been carefully evaluated. Last but not least, a causal dependence relationship holds between the activity `Receive notice` and the activity `Discuss and evaluate patient`. The discussion and evaluation of the patient is indeed triggered (in an *Intake* process) by the request to start an intake procedure. Similarly for the causal dependence between the activities `Receive notice` and the activity `Determine treatment plan`.

Finally, among the norm-based dependence expressions, two historical dependences can be identified (between the pair `Assign intakers` and `Update pat. folder with first intaker information` and between the pair `Assign intakers` and `Update pat. folder with second intaker information`). Indeed, an intaker is allowed to report information in the patient folder only if she has been appointed to do it, i.e., a historical dependence relationship holds between the two activities (and, hence, the latter cannot occur before the former).

The additional information that the dependence expressions are able to provide, makes it clear to the new director that a dependence relationship holds between the activities `Assign intakers` and `Update pat. folder with first intaker information`, as well as the reason why they have to occur in that specific order. Hence, making explicit these dependences helps the new director to understand why the procedure has been designed as it is.

Table 2. Dependence expressions characterizing the *Intake* process.

Ontological dependences			
Law-of-nature	Hist(RECN,PN)	Hist(RPF,ANPF)	Hist(RPF,UPFMF)
	Hist(RPF,UPFFI)	Hist(RPF,UPFSI)	Hist(PC,HC)
	Cause(RN,AN)	Cause(AMF,RMF)	Cause(ATP,RTP)
Business Goal	Cause(RN,DEP)	Cause(RN,DTP)	Hist(DEP,DTP)
	Cooc(RTP,DTP)	Cooc(RTP,DEP)	
Norm	Hist(AI,UPFFI)	Hist(AI,UPFSI)	

5.2   Business Process Redesign

It is often the case that business process models need to be redesigned. This can be due to different reasons e.g., because the world, the organization or the procedure they describe changes, or for optimization reasons. Several approaches and techniques have been investigated in the BPM community in order to support business analysts in business process redesign (see e.g., [9,24]).

Let us assume, that the new director of the healthcare institute, in order to better understand the efficiency of her institute, has appointed a business analyst to analyze the processes carried out in the institute. By analyzing the process under the perspective of evaluating its cycle time, the business analyst notices that the process presents some bottlenecks. Indeed, the activities Print notice, Receive Payment Ticket and Receive medical file have a high average duration time (6, 12 and 24h, respectively). In the first case, the high duration time is due to the fact that only one printer is available in the institute, while in the second and in the third case this is due to the response time required by patients and medical doctors to pay the ticket and to provide the medical file, respectively. Moreover, although in the last case the request of the file from the doctor is optional, it is needed in 95% of the cases. This causes a high average process cycle time[3] $(= 53.4h)$. In order to solve the issue, the

---

[3] The computation of the average process cycle time is based on flow analysis [9] and depends on the structure of the process. In this case, the average time required for a process execution is given by the average time required by: (i) the sum of the time required by the activities in sequence before the first split AND gateway, which is, in turn, given by the sum of the average times of the activities in sequence $((1 + 1 + 1.2 + 6 + 0.5 + 2 + 0.5)h = 12.2h)$; (ii) the sum of the times required by the most costly branches of the two AND blocks, i.e., the one dealing with the optional request to the doctor of the medical file and the one related to the ticket payment receipt. The former is computed as the weighted (with the corresponding probabilities) average of the two alternative branches between the XOR split and the XOR join, (i.e., $((0.95 * (0.5 + 24 + 1.5)) + (0 * 0.05))h = 24.7h)$, while the second is the sum of the average cycle time of the activities Ask for ticket payment and Receive ticket payment, (i.e., $(0.5 + 12)h = 12.5h)$, respectively; and (iii) the time required by the last two activities (i.e., $(3 + 1)h = 4h)$. The average cycle time is hence $(12.2 + 24.7 + 12.5 + 4)h = 53.4h$.

institute director, at the suggestion of the business analyst, decides to redesign the process.

In order to reduce the overall cycle time of the procedure, the business analyst suggests to apply two business process behaviour heuristics: *parallelism* and *resequencing* [9]. While the first heuristic consists of evaluating what "can be executed in parallel", the second one consists of "moving the activities to more convenient places" [9]. According to the process re-design heuristics, the business analyst suggests to (i) parallelize the printing of the notice and the enrichment of the patient file up to the storing of the intaker assignments; (ii) anticipate the request of the payment to the patients and the request of the medical file to the doctor. Figure 3 shows the redesigned model. Such a redesign allows the healthcare institute to save about 16.5 h of average cycle time by reducing the cycle time from 53.4 to 36.9 h - as most of the flow related to the notice management and to the intaker assignment is actually in parallel with the costly time required for waiting for the medical file.

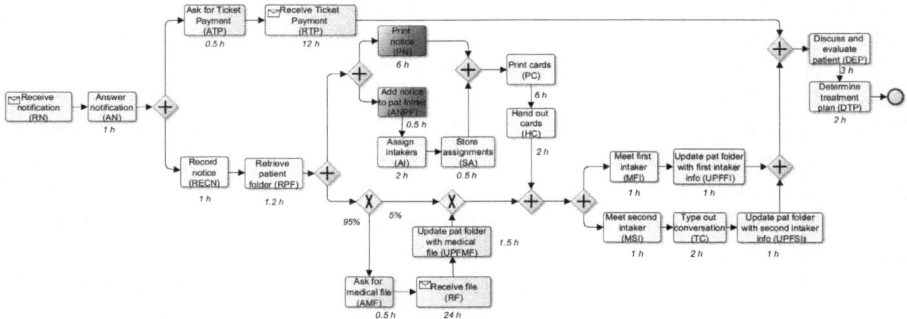

Fig. 3. *Intake* process redesigned according to the analyst's suggestions

However, by looking at the dependence expressions reported in Table 2, the business analyst can easily notice that, while anticipating the request of the medical file to the doctor and the ticket payment to the patient (depicted in green in the diagram) does not violate any of the identified dependences, this is not the case for the parallelization of the printing notice and the enrichment of the patient folder (marked in red). Indeed a historical dependence relationship holds between the activity Add notice to the patient folder and the activity Print notice, so that swapping them would result in an incorrect model.

Automated Check of Dependence Expression Enforcement. As reported in Sect. 4, the formalisation of dependence expressions in terms of LTL$_f$, allows us to take advantage of existing works (e.g., [7,16]) for the automated check of the enforcement of declarative properties or rules on procedural models, explanation of possible violations and repair actions. For instance, in the scenario described above, these techniques can be leveraged by the analyst to detect the inconsistency between the ontological dependence expressions and the redesigned process model, thus enabling the application of the only redesign heuristics that do not violate any dependence expression.

6   Related Work

We can roughly classify the literature related to this paper into three main groups: (i) works dealing with the analysis of business process model notations and its elements; (ii) works leveraging ontological analysis of business process modelling notations and its elements; and, finally, (iii) works combining declarative and procedural models.

Several papers in the literature focus on the analysis of the *elements* involved in business processes and business process modelling languages. Many of these works provide a comparison of different modelling notations [15,29] or develop metamodels of business process models across notations [14,19]. Other works, instead, take an ontological perspective to achieve the same goal. Indeed, some of them use ontologies for guiding the development of conceptual models and domain ontologies [4] or for semantically enriching business process models [12, 21,25], while others provide upper-level ontologies for business processes [22].

The second category of works leverages ontological analysis to deal with business process notations and business process model elements. Within this category, we can find works using the ontological analysis of business process elements (e.g., participants) across notations, such as [2]. In [26] the authors offer an ontological analysis of BPMN 2.0 elements and choreography diagram elements, respectively, with particular emphasis on the ontological characterization of BPMN events and activities. In [3] an effort towards a semantic foundation of the notion of *role* in the enterprise is provided. However, none of these works deals with the analysis of dependences among activities.

Indeed, although many efforts have been carried on so far in order to characterize ordering relationships between business process activities, an ontological analysis of these dependences has not been proposed yet. The analysis presented in this paper has been stimulated by philosophical and ontological papers like [5,10,17] which are strongly focused on defining and classifying ontological dependences, and where distinctions like weak vs rigid, ontological vs existential dependence are presented. Dependence as a schema is further discussed in [27] where an initial list of qualifications is also attempted (financial, practical, physiological, functional, ontological, logical and so on). Investigations on ordering relationships between activities are present also in the BPM community. An example is [8], where a definition of causal relation has been proposed as a sequence of events that can not be ordered in the opposite direction. Nevertheless, none of these works explicitly deals with ontological dependences in the context of business processes.

Several works combining declarative and procedural models have been recently investigated in the literature, some of which also provide automated support to deal with such a combination. Examples of the latter are works dealing with the automated discovery of hybrid process models [18,28], the automated check of a declarative formula on a procedural model [16], as well as the automated enforcement of the declarative component on the procedural one [7].

7 Conclusions

Existing business process modelling notations mainly focus on the representation of a specific kind of relationship between activities, that is their execution ordering within the control flow. However, the relationships between their activities of real-world processes are much richer and go beyond such a privileged relationship, covering relational constraints of different nature (e.g., ontological ones). In this paper we provided a characterisation of three ontological relationships (a.k.a. dependences) between business process activities: historical dependence, causal dependence, and goal-based co-occurrence. We introduced a language (for expressing them), made a proposal on how to incorporate them in business process models by adopting a hybrid approach, and showed their importance by discussing two application scenarios.

In the future, on the one hand, we would like to further investigate the ontological dependences between business process activities, by analysing the role and the ontological implications that business process participants (e.g., data objects, actors) have on the characterization of these ontological dependences; on the other hand, we are interested to extend our exploration of ontological relationships also to the relationships between activities and other types of business process participants.

Acknowledgments. This research has been partially carried out within the Euregio IPN12 KAOS, which is funded by the "European Region Tyrol-South Tyrol-Trentino" (EGTC) under the first call for basic research projects.

References

1. van der Aalst, W.M.P., Weijters, T., Maruster, L.: Workflow mining: discovering process models from event logs. IEEE Trans. Knowl. Data Eng. **16**(9), 1128–1142 (2004)
2. Adamo, G., Borgo, S., Di Francescomarino, C., Ghidini, C., Guarino, N., Sanfilippo, E.M.: Business processes and their participants: an ontological perspective. In: Esposito, F., Basili, R., Ferilli, S., Lisi, F. (eds.) AI*IA 2017 Advances in Artificial Intelligence. AI*IA 2017. Lecture Notes in Computer Science, vol. 10640, pp.215–228. Springer, Cham (2017). https://doi.org/10.1007/978-3-319-70169-1_16
3. Almeida, J.P.A., Guizzardi, G., Santos Jr., P.S.: Applying and extending a semantic foundation for role-related concepts in enterprise modelling. Enterp. Inf. Syst. **3**(3), 253–277 (2009)

4. Benevides, A.B., Guizzardi, G.: A model-based tool for conceptual modeling and domain ontology engineering in OntoUML. In: Filipe, J., Cordeiro, J. (eds.) ICEIS 2009. LNBIP, vol. 24, pp. 528–538. Springer, Heidelberg (2009). https://doi.org/10.1007/978-3-642-01347-8_44

5. Correia, F.: Ontological dependence. Philos. Compass **3**(5), 1013–1032 (2008)

6. De Giacomo, G., De Masellis, R., Montali, M.: Reasoning on LTL on finite traces: insensitivity to infiniteness. In: Proceedings of the 28th AAAI Conference on Artificial Intelligence, pp. 1027–1033. AAAI Press (2014)

7. De Masellis, R., Francescomarino, C.D., Ghidini, C., Laponin, A., Maggi, F.M.: Rule propagation: adapting procedural process models to declarative business rules. In: 21st IEEE International Enterprise Distributed Object Computing Conference, (EDOC 2017), pp. 165–174. IEEE Computer Society (2017)

8. Desel, J.: Validation of process models by construction of process nets. In: van der Aalst, W., Desel, J., Oberweis, A. (eds.) Business Process Management. LNCS, vol. 1806, pp. 110–128. Springer, Heidelberg (2000). https://doi.org/10.1007/3-540-45594-9_8

9. Dumas, M., Rosa, M.L., Mendling, J., Reijers, H.A.: Fundamentals of Business Process Management. Springer, Heidelberg (2013). https://doi.org/10.1007/978-3-662-56509-4

10. Fine, K.: Ontological dependence. In: Proceedings of the Aristotelian Society, vol. 95, pp. 269–290 (1994)

11. Galton, A.: States, processes and events, and the ontology of causal relations. In: Proceedings of the 7th International Conference on Ontology in Information Systems (FOIS 2012), Frontiers in Artificial Intelligence and Applications, vol. 239, pp. 279–292. IOS Press (2012)

12. Ghidini, C., Di Francescomarino, C., Rospocher, M., Tonella, P., Serafini, L.: Semantics-based aspect-oriented management of exceptional flows in business processes. IEEE Trans. Syst. Man Cybern. Part C (Applications and Reviews) **42**(1), 25–37 (2012)

13. Governatori, G., Rotolo, A.: Norm compliance in business process modeling. In: Dean, M., Hall, J., Rotolo, A., Tabet, S. (eds.) RuleML 2010. LNCS, vol. 6403, pp. 194–209. Springer, Heidelberg (2010). https://doi.org/10.1007/978-3-642-16289-3_17

14. Heidari, F., Loucopoulos, P., Brazier, F., Barjis, J.: A meta-meta-model for seven business process modeling languages. In: IEEE 15th Conference on Business Informatics (CBI) 2013, pp. 216–221. IEEE (2013)

15. List, B., Korherr, B.: An evaluation of conceptual business process modelling languages. In: Proceedings of the 2006 ACM symposium on Applied computing, pp. 1532–1539. ACM (2006)

16. Lohmann, N., Fahland, D.: Where did i go wrong? In: Sadiq, S., Soffer, P., Völzer, H. (eds.) BPM 2014. LNCS, vol. 8659, pp. 283–300. Springer, Cham (2014). https://doi.org/10.1007/978-3-319-10172-9_18

17. Lowe, E.J.: The Possibility of Metaphysics: Substance, Identity, and Time. Clarendon Press, Oxford (1998)

18. Maggi, F.M., Slaats, T., Reijers, H.A.: The automated discovery of hybrid processes. In: Sadiq, S., Soffer, P., Völzer, H. (eds.) BPM 2014. LNCS, vol. 8659, pp. 392–399. Springer, Cham (2014). https://doi.org/10.1007/978-3-319-10172-9_27

19. Mili, H., Tremblay, G., Jaoude, G.B., Lefebvre, É., Elabed, L., Boussaidi, G.E.: Business process modeling languages: sorting through the alphabet soup. ACM Comput. Surv. (CSUR) **43**(1), 4 (2010)

20. Moody, D.L.: The "Physics" of notations: toward a scientific basis for constructing visual notations in software engineering. IEEE Trans. Softw. Eng. **35**(6), 756–779 (2009)

21. Natschläger, C.: Towards a BPMN 2.0 ontology. In: Dijkman, R., Hofstetter, J., Koehler, J. (eds.) BPMN 2011. LNBIP, vol. 95, pp. 1–15. Springer, Heidelberg (2011). https://doi.org/10.1007/978-3-642-25160-3_1

22. Nicola, A.D., Lezoche, M., Missikoff, M.: An ontological approach to business process modeling. In: Proceedings of 3rd Indian International Conference on Artificial Intelligence, pp. 1794–1813 (2007)

23. Pesic, M., Schonenberg, H., van der Aalst, W.: DECLARE: full support for loosely-structured processes. In: Proceedings of the 11th IEEE International Enterprise Distributed Object Computing Conference (EDOC 2007), pp. 287–300. IEEE Computer Society (2007)

24. Reijers, H.A., Liman Mansar, S.: Best practices in business process redesign: an overview and qualitative evaluation of successful redesign heuristics. Omega **33**(4), 283–306 (2005)

25. Rospocher, M., Ghidini, C., Serafini, L.: An ontology for the business process modelling notation. In: Proceedings of 8th International Conference on Formal Ontology in Information Systems (FOIS 2014). Frontiers in Artificial Intelligence and Applications, vol. 267, pp. 133–146. IOS Press (2014)

26. Sanfilippo, E.M., Borgo, S., Masolo, C.: Events and activities: is there an ontology behind BPMN? In: Proceedings of 8th International Conference on Formal Ontology in Information Systems (FOIS 2014). Frontiers in Artificial Intelligence and Applications, vol. 267, pp. 147–156. IOS Press (2014)

27. Simons, P.: Parts: a Study in Ontology. Clarendon Press, Oxford (1987)

28. Smedt, J.D., Weerdt, J.D., Vanthienen, J.: Fusion miner: process discovery for mixed-paradigm models. Decis. Support Syst. **77**, 123–136 (2015)

29. Söderström, E., Andersson, B., Johannesson, P., Perjons, E., Wangler, B.: Towards a framework for comparing process modelling languages. In: Pidduck, A.B., Ozsu, M.T., Mylopoulos, J., Woo, C.C. (eds.) CAiSE 2002. LNCS, vol. 2348, pp. 600–611. Springer, Heidelberg (2002). https://doi.org/10.1007/3-540-47961-9_41

30. Weske, M.: Business Process Management. Concepts, Languages, Architectures. Springer, Heidelberg (2012). https://doi.org/10.1007/978-3-642-28616-2

Enriched Modeling and Reasoning on Business Processes with Ontologies and Answer Set Programming

Laura Giordano and Daniele Theseider Dupré[✉]

DISIT, Università del Piemonte Orientale, Alessandria, Italy
laura.giordano@uniupo.it, dtd@di.unipmn.it

Abstract. Domain ontologies may provide the proper level of abstraction in modeling semantic constraints and business rules in BPM; in fact, ontologies are intended to define terminologies to be shared within and across organizations and reused in different applications. In this paper we show how Answer Set Programming (ASP), a powerful framework for declarative problem solving, can accommodate for domain ontologies in modeling and reasoning about Business Processes, especially for process verification. Description Logics (DLs) provide the formal counterpart of ontologies, and in our approach knowledge on the process domain is expressed in a low-complexity DL. Terms from the ontology can be used in embedding business rules in the model as well as in expressing constraints that should be verified to achieve compliance by design. Causal rules for reasoning on side-effects of activities in the process domain can be derived, based on knowledge expressed in the DL. We show how ASP can accommodate them, relying on a reasoning about actions and change approach, for process analysis, and, in particular, for verifying formulas in temporal logic.

1 Introduction

In this paper we show how we can accommodate in Answer Set Programming (ASP) several sources of knowledge for reasoning on Business Processes, in particular, for verification purposes, i.e. in order to ensure process compliance. A process model expressed in a standard business process modeling language is enriched with **domain knowledge**, in particular, ontological knowledge describing terms used:

1. in conditions on sequence flow, in particular, conditions in data-based split gateways;
2. in *semantic constraints* on the process, i.e., constraints that express "dependencies such as ordering and temporal relations between activities, incompatibilities, and existence dependencies" [31].

As discussed in [31], semantic constraints are a subset of *business rules*; while this term in widely used in the BPM context, it is a broad one that comprises several types of knowledge about a business domain. Both [36,37] present

© Springer Nature Switzerland AG 2018
M. Weske et al. (Eds.): BPM Forum 2018, LNBIP 329, pp. 71–88, 2018.
https://doi.org/10.1007/978-3-319-98651-7_5

attempts at classifying types of business rules, and the semantic constraints considered in [31] are stated to be *action assertions* in terms of the classification in [36]; in terms of [32,37], semantic constraints can be intended as integrity rules, while derivation and reaction rules are suitable to be embedded in the process description.

As [31] points out, semantic constraints abstract from the way some fact about the case at hand may be actually represented, or computed from stored data, in the process implementation. This is why we believe that ontological knowledge is especially suited for expressing them. In fact, the very idea of the Semantic Web and, in general, of terminological knowledge bases, includes the fact that terminological knowledge about a domain can be shared and reused in several applications. Business Process Management applications can therefore reuse existing terminologies about a whole domain (like the well-known SNOMED-CT medical terminology [29]), and an organization can define its own terminology to be reused in several applications, including management of its own processes. Domain ontologies are also believed to facilitate shared understanding of the process domain across team members [34].

In [31] it is also pointed out that compliance with semantic constraints may be checked at design time (compliance by design [35]), even though not all exceptional situations and process changes (to deal with exceptions) may be considered in the process model, to avoid it becoming too complicated in order to be readable. Therefore, it may be necessary to reason about such constraints at runtime, as part of execution support; i.e., at design time compliance is checked under the assumption that no exception occurs, while at runtime the actual exceptions and process changes occurring in the case at hand are considered; conflicts with semantic constraints should be pointed out as well as possible ways for restoring consistency with them. Semantic constraints are also useful in providing intelligible feedback to users and in supporting traceability, e.g., in order to point out whether and where, in a given process execution, they were violated.

In the work presented in this paper we incorporate contributions from Artificial Intelligence (Logic-based Knowledge Representation and Reasoning) and Formal Methods:

– **Modeling and reasoning based on description logics** [5]. Terminological knowledge has been identified, starting from the 1980's, as a form of knowledge which can be expressed in suitable sublanguages of first-order logics, *description logics* (DLs), as well as being useful in formalizing definitions of the terms used in several domains (as pointed out earlier). While full first-order logic is undecidable, DLs offer a trade-off between expressiveness and computational complexity of reasoning, some of them enjoying low complexity while still being able to describe wide terminologies (see, e.g., the already mentioned SNOMED-CT terminology which can be expressed in $\mathcal{EL}$ [3]). As a result, description logics have been chosen as the basis for the Semantic Web, and, in particular, the Web Ontology Language (OWL).

– **Reasoning about action and change** [21], where a domain is described in terms of **fluents**, i.e., propositions whose value can change, possible **actions**

which have preconditions, direct effects in terms of fluents, and static and dynamic **causal laws** which model dependencies between truth values of fluents, or changes in such truth values.
- **Formal verification** based on temporal logics [7], such as LTL.

For the purposes of this paper, all of them (at least, limiting to low-complexity description logics) can be integrated in Answer Set Programming (ASP) [18], a powerful framework for declarative problem solving which combines significant modeling capabilities with efficient solving, relying on inference techniques that include the ones used in SAT solvers.

In fact, in our previous work we showed the following:

1. ASP (which has been used for reasoning about action already in [8,15,16,26]) can be used for verification (with Bounded Model Checking) of properties, expressed in an extension of Linear Temporal Logic, of an action domain modeled in terms of fluents, action laws providing direct effects of actions, and causal laws [24].
2. The previous framework can be used for reasoning on business processes, in particular, for verifying process properties in temporal logic; the process can be modeled in terms of the widely adopted workflow-like languages (as well as in declarative ones). The approach is described in [22], where, relying on Constraint Answer Set Programming [19], the framework in [24] is extended to deal with conditions on numerical variables, used in the process model in data-based conditions on exclusive splits, and in the formulae to be verified. Process activities correspond to actions, and fluents are used in modeling the enabling of activities (according to the workflow model) as well as further pre/post conditions for activities, expressed in terms of process variables and further background fluents (in particular, postconditions correspond to *annotations* proposed already in [28]).
3. Reasoning about actions performed in ASP can rely on domain knowledge in a low-complexity DL [23]. More precisely, axioms in the DL describe static knowledge on a domain, e.g.: someone that teaches a university course is a lecturer. Causal laws should be associated with such knowledge to control which fluents may change as side effects of other changes, in order for the axiom to still hold, after an action whose direct effects are explicitly stated. Extra knowledge may be necessary to avoid all potential ways for restoring truth of the axiom; e.g., starting from the situation where John teaches university course CS101 (and is then inferred to be a lecturer, according to DL knowledge), if an action (such as John retiring or being fired) has the direct effect he is no longer a lecturer, we would like to infer as a side effect that he does no longer teach CS101 (nor any other course he was teaching), without considering the scenario where CS101 ceases to be a university course which would, in principle, be another way of restoring consistency.

Building on these contributions, in this paper we propose an approach to process modeling and semantic analysis that is able to exploit terminological knowledge in relying process activities to semantic constraints, via the definition of effects and preconditions of activities, and domain knowledge that relates such effects to the terms used in semantic constraints. The proposed approach exploits for business process verification the Bounded Model Checking verification methodology in ASP developed in [24].

We believe that this can indeed enrich process modeling and analysis in the BPM field, given that it provides expressive modeling at the semantic level and relies on the power of ASP solvers for efficient inference.

In the next section we summarize the sources of knowledge of our approach, and how they are expressed. In particular, in Sect. 2.1 we describe how the terminological domain knowledge base is represented, and in Sect. 2.2 we discuss how such a knowledge base can be used in reasoning about action and change. In Sect. 3 we describe how a process model can be described in terms of the framework in the preceding section. Section 4 is devoted to explaining how the model can be encoded in ASP, and how it can be used for process verification. We finally discuss the properties of our contribution especially in comparison with related work in the literature.

2   Sources of Knowledge

As sources of knowledge we consider the following ones.

- A **domain knowledge base** describing terms in the process domain: unary predicates (*classes*) describing entities of the domain, and binary predicates, called *roles* or *properties*, describing relations among domain entities. The knowledge base is formalized as a set of description logic axioms and causal rules, detailed in Sects. 2.1 and 2.2. At least some of the class predicates and properties are *fluents*, i.e., may change their truth values as effect of the process activities.
- A model for the **sequence flow** of the process is given, using conventional gateways. In particular, we refer to BPMN, and in the following we limit our consideration to models using activities, exclusive and parallel gateways (i.e., XOR splits and joins, and AND split and joins).
  Following BPMN, *conditions* on data can be attached to the sequence flow, out of gateways, in particular, exclusive gateways, thus providing *data-based exclusive gateways*. BPMN allows for specifying (in the expressionLanguage attribute of a model) a language to be used for expressing such data-based conditions. In this paper we do not detail the specification of one such language, but we intend that data-based condition expressions may use terms from the domain knowledge base.
- **Data objects** in the process and their **states**, which are also part of the BPMN standard, to model, e.g., an "order" whose set of states includes, for example, "pending" and "confirmed". The domain knowledge base may

mention such data objects and relate them to other entities in the process domain.

– **Pre- and postconditions** for activities. Postconditions are used to model the direct effects of activities in terms of the process domain. Postconditions include state changes for data objects, or, more generally, the case where a BPMN data object is output of an activity, i.e., the activity creates or writes the data object (represented as an output data association in BPMN). Similarly, preconditions include the ones that are represented in the BPMN process model by input sets for activities (input data associations).

2.1 The Domain Knowledge Base

We consider, as in [23], domain knowledge bases expressed in a fragment of the description logic $\mathcal{EL}^{++}$ [3]. The fragment, $\mathcal{EL}^{\perp}$, includes the concept $\perp$ (the empty concept, which is false for all individuals) as well as nominals, i.e., concepts corresponding to single individuals.

As for other description logics, the language of $\mathcal{EL}^{\perp}$ is based on a set N_C of concept names (class names), a set N_R of role names (names for properties, i.e., binary relations) and a set N_I of individual names. A concept in $\mathcal{EL}^{\perp}$ is defined as follows:

$$C := A \mid \top \mid \perp \mid C \sqcap C \mid \exists r.C \mid \{a\}$$

where $A \in N_C$, $r \in N_R$ and $a \in N_I$. That is:

– Concept expressions include class names (named concepts) and the concepts "true" and "false".
– A concept can be an intersection ($\sqcap$) of concepts (i.e., named concepts or concept expressions).
– A concept can be built from a role name r and a concept C using an *existential restriction*: the instances of concept $\exists r.C$ are the individuals x which are in relation r with some member y of the concept C.
– A concept can be the *nominal* $\{a\}$, i.e., the concept of "being a".

A knowledge base in $\mathcal{EL}^{\perp}$ is a pair $(\mathcal{T}, \mathcal{A})$, where:

– $\mathcal{T}$ (a *TBox*, i.e., the *terminological* part) is a finite set of concept inclusions $C_1 \sqsubseteq C_2$, where C_1 and C_2 are concepts,
– $\mathcal{A}$ (an *ABox*, the *assertional* part) is a set of assertions of the form $C(a)$ and $r(a, b)$, where C is a concept, $r \in N_R$ and $a, b \in N_I$.

The TBox can be expressed in a normal form [4] where axioms only have the forms: $C_1 \sqsubseteq D$, $C_1 \sqcap C_2 \sqsubseteq D$, $C_1 \sqsubseteq \exists r.C_2$, $\exists r.C_2 \sqsubseteq D$, where C_1, C_2 are from BC_{KB}, i.e., the set of concepts containing $\top$, all the named concepts occurring in KB and all nominals $\{a\}$, for any individual name a occurring in KB; and D is in $BC_{KB} \cup \{\perp\}$.

The semantics of $\mathcal{EL}^{\perp}$ is defined in the usual way for description logics, based on a domain (a set) Δ, an interpretation of individuals as elements of Δ, of concept names as subsets of Δ, of role names as binary relations on Δ. The interpretation of concept expressions is defined formalizing the description given above for the meaning of such expressions. An concept inclusion $C_1 \sqsubseteq C_2$ is satisfied in an interpretation if the interpretation of C_1 is a subset of the interpretation of C_2 (see [3, 23] for the formal definitions).

Examples of concepts are:

- $\exists Teaches.Course$, whose instances are the domain elements who teach a course;
- $\exists Teaches.\{cs101\}$, the ones who teach the individual course $cs101$;
- $UndergraduateCourse \sqcap ComputerScienceCourse$, the concept of undergraduate courses in computer science, which is expressed as the intersection of undergraduate courses and computer science courses.

Examples of concept inclusions are:

- $\exists Teaches.Course \sqsubseteq Lecturer$, which states that the ones who teach some course are lecturers;
- $Course \sqcap \exists HasSubject.ComputerScienceSubject \sqsubseteq ComputerScienceCourse$, which states that a course, which has as subject a computer science subject, is a computer science course. Adding the inverse inclusion would provide a definition of $ComputerScienceCourse$.

2.2   Reasoning About Actions with Terminological Knowledge

Given a domain which is modeled with a knowledge base in a description logic, as above, when reasoning about a process in the domain involving actions and changes, we will assume that the Tbox axioms do not change and must be satisfied during all the process execution, even though, in the long term, it could be appropriate to allow for the knowledge base to evolve with new concepts and new axioms about them, while other axioms may cease to hold, or may be modified, e.g., in order to provide coverage of cases that were not considered before. Presumably, the process model itself should change as well, and automated reasoning may support process redesign, but we do not address the issue in this paper and assume that both the process model and the Tbox do not change.

Of course, we do consider that Abox assertions may change as a result of actions (they are *fluents*). We summarize in the following the way reasoning about such actions and changes can be defined [23] to take into account background knowledge about the domain expressed in an $\mathcal{EL}^\perp$ Tbox. The axioms in the Tbox can be regarded as *state constraints*, the term used in the literature in reasoning about actions and change to describe conditions that must hold in all states.

The language we consider for reasoning about action and change involves predicate symbols and constants. Such symbols include the concept names, role names and individual names occurring in the $\mathcal{EL}^\perp$ domain knowledge base.

The *fluents* $\mathcal{F}$ are ground atomic propositions $p(a_1, \ldots, a_k)$ where p is a predicate symbol and $a_1, \ldots, a_n$ are constants.

A *fluent literal* l is a fluent f or its explicit negation $-f$. Two literals f and $-f$ are the *complement* of each other. We denote by Lit the set of fluent literals.

If a concept $\exists r.C$ occurs in the KB, the predicate names in the action theory include a name $\exists r.C$, so that, for a individual name a, the fluent literals $(\exists r.C)(a)$ and $-(\exists r.C)(a)$ belong to Lit.

A *state* S is a set of literals in Lit. A state S is *consistent* if it is not the case that both a literal and its complement belong to S. A state S is *complete* if for any fluent literal l, S contains l or its complement.

For describing an action theory, laws are introduced in a notation in the line of the literature of reasoning about actions and change [8, 16, 26]. *Action laws* describe the direct effects of actions. They have the form:

$$\alpha \text{ causes } \phi \text{ if } \psi_1 \text{ after } \psi_2$$

meaning that the execution of action α in a state in which ψ_2 holds causes ϕ to hold in the new state as a direct effect, if ψ_1 holds in the new state as well. The action name α corresponds to an activity in a process model, ϕ is a literal in Lit and $\psi_i = L_1 \wedge \ldots \wedge L_m, not\ L_{m+1} \wedge \ldots \wedge not\ L_n$ is a conjunction of literals $L_i \in Lit$ or their default negations. The informal meaning of *default negation* in $not\ L_j$ is that "L_j is not believed", its formal semantics is the stable model semantics [20].

The action name can have parameters also occurring in ϕ, ψ_1 and ψ_2, and a parametric action law is a shorthand for all its instances with individual names; the same applied to other types of laws described below. An example (instance) of action law is:

$$retire(john) \text{ causes } - Lecturer(john)$$

Non-deterministic effects of actions can be defined using default negation in the body of action laws. For instance, after flipping a coin, the result may be head or not:

$$flip \text{ causes } head \text{ if } not\ - head$$
$$flip \text{ causes } - head \text{ if } not\ head$$

Causal laws describe indirect effects of actions. They have the form:

$$\text{caused } \phi \text{ if } \psi_1 \text{ after } \psi_2$$

meaning that ψ_1 causes ϕ to hold whenever ψ_2 holds in the previous state; ϕ is a literal in Lit and ψ_1 and ψ_2 are as in action laws. If the condition ψ_2 is $\top$, the causal law is said to be *static*, since it only involves conditions on a single state, and the **after** part is omitted.

An example causal law, that, as we shall see, could be associated with the Tbox axiom $\exists Teaches.Course \sqsubseteq Lecturer$, is:

$$\text{caused } Lecturer(x) \text{ if } Teaches(x, y) \wedge Course(y)$$

Precondition laws describe the executability conditions of actions. They have the form: α **executable if** ψ, meaning that the execution of action α is possible in a state where the precondition ψ holds; α is an action name and ψ is a conjunction of literals or default negations of literals. An example is: *retire(x)* **executable if** *aged(x)*.

The *constraints* define conditions that must be satisfied by all states. They have the form: $\perp$ **if** ψ, meaning that any state in which ψ holds is inconsistent.

Initial state laws are needed to introduce conditions that have to hold in the initial state. They have the form: **Init** ϕ **if** ψ. When $\phi = \perp$, we get the a constraint on the initial state **Init** $\perp$ **if** ψ.

Most fluents are intended to be *frame* fluents, i.e., their truth value persists across action occurrences. For all such fluents p, the following causal laws, said *persistency laws*, are introduced:

$$\textbf{caused } p \textbf{ if } not \; -p \textbf{ after } p$$
$$\textbf{caused } -p \textbf{ if } not \; p \textbf{ after } -p$$

meaning that, if p holds in a state, then p will hold in the next state, unless its negation $-p$ is caused to hold (and similarly for $-p$). Persistency of a fluent is blocked by the execution of an action which causes the value of the fluent to change, or by a nondeterministic action which may cause it to change.

Persistency laws are not provided for literals such as $(\exists r.C)(a)$; their value in a state is rather derived, using causal laws, from the one of literal with "simple" predicate names, i.e., $(\exists r.B)(x)$ is caused if $r(x,y) \wedge B(y)$.

Initial state laws that correspond to what is known about the initial state may incompletely specify it. As we want to reason about all the possible complete states, the laws:

$$\textbf{Init } p \textbf{ if } not \; -p$$
$$\textbf{Init } -p \textbf{ if } not \; p$$

for completing the initial state are introduced for all "simple" literals p.

In [23] a **semantics** is defined for action execution. Given a state (a set of literals) S which is consistent and complete (i.e., it contains either l or $-l$ for all fluent literals), such a semantics defines which are the possible resulting states if an action α is executed in S, and is based on the answer set semantics [18].

We assume to start action execution in a state which satisfies the Tbox $\mathcal{T}$; however, in general there is no guarantee that, if an action α is applied to a state satisfying $\mathcal{T}$, the state that can result, according to the semantics, will still satisfy $\mathcal{T}$.

However, suitable causal laws can be associated with (normalized) axioms in $\mathcal{T}$ in order to guarantee this[1] [23]; here we describe part of them. For inclusions $A \sqsubseteq B$, two causal laws are needed:

$$\textbf{caused } B(x) \textbf{ if } A(x) \text{ and } \textbf{caused } - A(x) \textbf{ if } - B(x)$$

For an axiom $\exists r.B \sqsubseteq A$, the laws:

$$\textbf{caused } A(x) \textbf{ if } (\exists r.B)(x)$$
$$\textbf{caused } -(\exists r.B)(x) \textbf{ if } - A(x)$$

and at least one of:

$$\textbf{caused } -r(x, y) \textbf{ if } - A(x) \land B(y)$$
$$\textbf{caused } -B(y) \textbf{ if } - A(x) \land r(x, y)$$

should be introduced.

For example, an axiom $\exists approved\_by.examiner \sqsubseteq approved$ relative to insurance claim processing, has the associated causal law:

$$\textbf{caused } approved(x) \textbf{ if } (\exists approved\_by.examiner)(x)$$

where $(\exists approved\_by.examiner)(x)$ is in turn caused, if $approved\_by(x, y)$ and $examiner(y)$). If we admit that the claim, after being approved by an examiner, can be made $-approved$ by a manager, the causal law:

$$\textbf{caused } - approved\_by(x, y) \textbf{ if } - approved(x) \land examiner(y)$$

is introduced, while the other possible causal law is not, because we do not expect $examiner(y)$ to become false as a side effect of $approved\_by(x, y)$ becoming false.

There is an option also for the case of an axiom $A \sqcap B \sqsubseteq D$; besides the law $\textbf{caused } D(x) \textbf{ if } A(x) \land B(x)$, at least one of:

$$\textbf{caused } - A(x) \textbf{ if } - D(x) \land B(x)$$
$$\textbf{caused } - B(x) \textbf{ if } - D(x) \land A(x)$$

should be introduced.

The presence of such options requires further pieces of domain knowledge, besides the axioms in $\mathcal{T}$. The choice of causal rules to be discarded should in general be made for each single axiom, while in some cases it can be derived from more general knowledge. In the literature about ontologies, and, in particular, Temporal Description Logics, a distinction is introduced between *rigid* and *temporal* concepts and roles [1], i.e., the ones that are supposed not to change their truth values across time, and the ones that may change. If such a distinction is present, it can be used to discard optional causal rules: if a concept or role

---

[1] As a consequence of the introduction of the causal laws for the axioms in $\mathcal{T}$, there is no need to exploit a DL reasoner, as each state is guaranteed to satisfy $\mathcal{T}$.

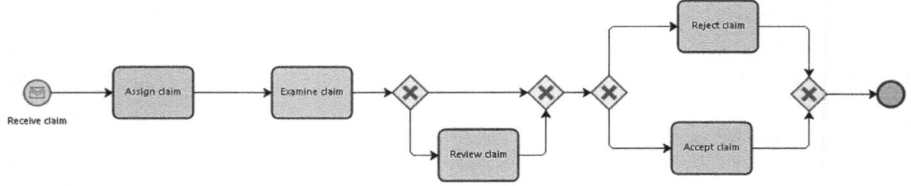

Fig. 1. Example process model

is identified as rigid, optional causal rules where concept or role occurs in the head (the "caused" part) will not be included. Transformation rules, a type of business rules identified in [32,37], can also represent a source of knowledge for selecting optional causal rules.

As detailed in [23], the representation of an action domain with terminological knowledge can be encoded in ASP, and an ASP solver can be used to determine, e.g., whether a given literal holds in some of the resulting states, or in all the resulting states, after executing a given sequence of actions. In Sect. 4 we present such an encoding and show that it can be combined with the ASP representation of the sequence flow of a process model, and an ASP representation of Bounded Model Checking, in order to verify semantic constraints.

Before that, in the next section we show how an action domain description can be derived, in a semi-automated translation, from a process model described using the basic elements of BPMN.

3   Process Models as Action Domains

Consider a simple process model for insurance claim processing whose control flow is described in Fig. 1 (additional knowledge is not shown since only part of it can be represented in BPMN). In this model, a claim is assigned to a claims examiner, who provides a (preliminar) acceptance or rejection, and then possibly reviewed by a claims manager. We do not detail the accept/reject final part in terms of sending letters or performing payment.

All activities refer to a data object *Claim*, which is output of the start event *Receive claim* and is both input and output of all the other activities.

The activity *Assign claim* also has as output the examiner who had the claim assigned and the manager who should possibly review the claim. Examiner and manager are input to the activities executed by them (alternatively, swim lanes could be used to represent actors in the process).

In general, in the representation of activities as actions in the action domain, a choice should be made on which parameters are introduced for the action. As a default, if an activity a has a data object as input, it will have it as a parameter. Therefore, all activities in the process have as parameters a claim identifier, and the person executing the activity.

When executing the process, it is of course necessary to assign specific values to data objects (the claim identifier, the examiner and manager, as well as other variables that may appear in the model). When reasoning about the process, considering all possible values that a data object can assume could be unnecessary (other than, of course, source of intractability or undecidability). In the example model, the actual value of the claim identifier and the actual names of the examiner and manager are not supposed to influence process execution (as we shall see, they do not occur in the data-based conditions), and are therefore irrelevant.

Then, when an object is output of an activity, we represent its value with an individual name with the only constraint that it should be different from other names. We will then consider the parametric activities to be instantiated with such parameters. This can be considered as a default in an automatic translation, but, in general, such a translation can only be semi-automated, given that the choice of a suitable abstraction on data is essential for the model to be useful. As discussed before, ontologies can help to this purpose: in the example considered here, suppose there are two classes of examiners, expert ones and in training, and that process execution depends on the class of the examiner, but not on who actually the examiner is. Then, in the model, two alternative, nondeterministic postconditions in terms of examiner type can be considered, rather than all possible individual names.

In the example we will use as values the names *claim, examiner, manager* of the data objects themselves. To avoid redundancy, we remove "claim" from the name of the activity. The action instances that are considered in the action domain corresponding to the process model are then:

$assign(claim)$
$examine(examiner, claim)$
$review(manager, claim)$
$reject(examiner, claim)$
$accept(examiner, claim)$

The control flow of the process model can be represented with action laws and precondition laws resulting from an automated translation similar to the one described in [22] (Appendix A) for a subset of YAWL, analogous to the subset of BPMN used in this paper. In the following we use the terms "activity" (in the process model) and "action" (in the action domain) interchangeably. Fluents are introduced to represent the *enabling* of activities, which is a precondition for the action. In case of two activities in sequence, execution of the first one disables itself, and enables the next one. With a parallel split, all outgoing flows are enabled, and with parallel join, enabling is necessary from all incoming flows. With nondeterministic exclusive split, the pattern for nondeterministic actions is used.

Data-based conditions for exclusive splits are the most interesting case for the approach proposed in this paper, since, as pointed out in the introduction, they are the place in the model where terms from the domain knowledge (in particular, concept names and role names applied to data objects) can be conveniently used. For the model in Fig. 1, we suppose that the condition for reviewing a claim it that it is approved by the examiner and the customer is suspect of being a fraudster (in a variation of the example, another sufficient condition could be that the examiner is in training). The condition can be expressed as *PossiblyFraudolentClaim(claim)* where the concept is defined in the domain knowledge base as *Claim ⊓ ∃HasCustomer.SuspectFraudster*. How a customer is actually suspected to be a fraudster (also due to previous claim history) could indeed be one of the cases mentioned in the introduction, where in the model we want to abstract from the way this is explicitly stored or computed[2]. Notice that, given that the initial state is made complete (see Sect. 2.2), the possible complete initial states will contain either *PossiblyFraudolentClaim(claim)* or its complement.

Further action laws state that:

- *examine(examiner, claim)* has an effect *examined(claim)* and a nondeterministic effect *approved_by(examiner, claim)* or *−approved_by(examiner, claim)*
- *review(manager, claim)* has a nondeterministic effect *approved(claim)* or *−approved(claim)*.

The causal laws in Sect. 2.2, associated with the domain knowledge axiom *∃approved_by.examiner ⊑ approved*, imply that if the claim is approved by the examiner and does not undergo review, it will remain approved; while if it is made not approved by the manager's review, it will no longer be considered as approved by the examiner.

The second exclusive split is (obviously) conditioned on *approved(claim)*.

4   ASP Representation

An action domain, including the one derived from a process model as described in the previous section, can be represented in ASP as follows [23].

States are represented as integers, starting with the initial state 0. The predicate *occurs(Action, State)* represents the fact that *Action* occurs in *State*; occurrence of exactly one action in each state must be imposed:

$$-occurs(A, S) \leftarrow occurs(A1, S), action(A), action(A1), A{\neq} A1, state(S).$$
$$occurs(A, S) \leftarrow not\ -occurs(A, S), action(A), state(S).$$

(in state-of-the art ASP solvers [18], this can also be expressed with *choice rules*, whose syntax we do not introduce here).

In order to represent the fact that a literal holds in a state, we use different predicates:

- *holds_inst(Concept, Name, State)* represents the fact that an assertions of the form $C(a)$ holds in a state.
- *holds_triple(Role, NameA, NameB, State)* is used for role assertions $r(a, b)$

---

[2] It could be modeled separately in a decision model, an issue we do not address in this paper.

– $holds(Fluent, State)$ is used for other fluents (for an action domain derived from a process model, they are used to model control flow).

An *action law*:

$$\alpha \textbf{ causes } L_0 \textbf{ if } \psi_1 \textbf{ after } \psi_2$$

where $\psi_1 = L_1 \wedge \ldots \wedge L_m, not\ L_{m+1} \wedge \ldots \wedge not\ L_n$ and $\psi_2 = L_1' \wedge \ldots \wedge L_m', not\ L_{m+1}' \wedge \ldots \wedge not\ L_n'$ is translated to:

$$h_0 \leftarrow state(S), S' = S + 1, occurs(a, S), h_1 \ldots h_m, not\ h_{m+1} \ldots not\ h_n,$$
$$h_1' \ldots h_m', not\ h_{m+1}' \ldots not\ h_n'$$

where $h_0 = (-)holds\_inst(C_0, a_0, S')$ if $L_0 = (-)C_0(a_0)$, $h_0 = (-)holds\_triple$ (r_0, a_0, b_0, S'), if $L_0 = (-)r_0(a_0, b_0)$, and $h_0 = (-)holds(p(a_1, .., a_n), S')$ if $L_0 = (-)\ p(a_1, .., a_n)$ and similarly for the h_i's and h_j''s, using S' for the h_i's and S for the h_j''s; where C_0 in $holds\_inst$ stands for the ground term representing C_0, and similarly for $p(a_1, .., a_n)$.

Other laws can be translated to ASP in a similar way.

In [24] we showed (for a variant of the action language used in this paper, which can be similarly encoded in ASP) that, given an action domain, temporal properties of the domain, expressed in Dynamic Linear Time Temporal Logic [27], an extension of Linear Time Temporal Logic [7], can be verified in ASP in a Bounded Model Checking (BMC [9]) approach. The ASP encoding in [24] is suitable for verifying systems with infinite computations which can be finitely represented (with a loop back to a previously reached state). The approach relies on the definition of a predicate $sat(T\_alpha, S)$, where T_alpha is a term representing a temporal logic formula α, and S is a state, which corresponds to the fact that α holds in S. The predicate can be defined inductively on the structure of α.

In [22] we showed how the approach can be adapted to the verification of properties of finite executions of a business process model. Model checking should either find a counterexample for the formula to be verified (an interpretation falsifying the formula) or ensure that no counterexample exists. BMC is in general a partial decision procedure for model checking: it considers executions of bounded length, iteratively increasing the bound; if no model exists, in general the procedure would not stop. There are, however, cases where a *completeness threshold* can be identified (a value such that, if a counterexample exists, it can be found using such a value as bound). An obvious case is the one where the workflow of a business process model is loop-free. Appendix B of [22] reports results that demonstrate the scalability of the approach, also considering non-loop-free workflows. Processes with up to 200 activities (a size which is in line with the one of real-world processes in [17]) and run length of more than 100 activities are considered. Properties in LTL are verified, while DLTL can be useful for the declarative specification of process models.

The same approach[3] can be used for verifying LTL properties of action domains in this paper, where LTL formulae can be built from fluents, including

---

[3] The work in [22] allows for conditions on numerical data – e.g., the piece number in an order is larger than 50000 – to be used in the model and in the formulae to

assertions in the language of domain knowledge. The analysis is performed on the finite domain represented by the set of constants in the ASP encoding. This is without loss of generality as regards the domain knowledge, given that it is expressed in $\mathcal{EL}^{\perp}$; but it relies on the assumption that the domains for data objects are assumed to be finite.

As an example, the formula:

$$\Box(examined(claim) \wedge \neg approved(claim) \to \neg \Diamond approved(claim))$$

corresponds to the property that an examined claim which is not approved cannot become approved. In the model described in Sect. 3, it indeed holds, because the claim is reviewed only if it was approved by the examiner (and the customer is suspected to be fraudolent), while if was not approved by the examiner, it does not undergo review and its approval is not modified. The formula can be verified to hold using the approach described above.

We can observe that the (grounding of the) ASP program has polynomial size in the size of the input (ontology and business process). More precisely, let n be the size of the ontology, m the size of the business process, d the size of data domains and annotations, f the size of the formula to be checked by bounded model checking, and k (a constant) the length of the sequence searched for in a BMC verification of a formula (i.e. the number of states). The size of the BP encoding is $O(m \times d \times k)$, while the size of the action theory and BMC encoding is $O((n^2 + f) \times d \times k)$. $O(n^2)$ is an upper bound on the size of causal laws, taking into consideration the number of possible contrapositives of $\mathcal{EL}^{\perp}$ inclusions. From this observation and the fact that the final ASP encoding is a normal (non disjunctive) logic program, it follows that checking satisfiability of a temporal formula over a BP specification is in NP.

5   Conclusion and Related Work

In the paper we described how domain knowledge in the form of ontologies can be accommodated in modeling and reasoning about business processes in Answer Set Programming. We build on contributions in our previous work [22–24], but their combination for the verification of BPMN process models enriched with domain knowledge is novel. As a reasoning task, we emphasized verification of compliance by design, but other reasoning tasks can be accommodated as well. Consider, for example, compliance at runtime which should take into account a specific partial execution (whose events are given, up to a current time), but also exceptional situations occurring in the case at hand (exception not necessarily considered in the general process model). The model description in ASP

---

be verified. In order to deal with them, without considering all individual values in the – finite but large – numerical domain, it relies on Constraint ASP [19]. In this paper we do not consider this feature, which can however be integrated with the ones addressed here, and would provide another form of abstraction, complementary to the use of ontologies.

is modular, elaboration tolerant, and can easily accommodate, e.g., for additional actions with their enabling conditions (not necessarily related to the basic workflow structure).

Our contribution is related to several ones in the literature.

Ly et al. in [31] provide thorough motivations for the use of semantic constraints in BPM; first-order predicate logic is used as a language for expressing such constraints (while also mentioning description logics as a suitable option) but the paper does not describe the use of automated reasoning based on such a description of semantic constraints in logic. Actually, as pointed out in [31], ontological modeling and reasoning can also be useful to relate specific activities to abstract classes of activities, such as, in the medical domain, "invasive procedures", e.g., to ensure compliance of processes with the constraint that the patient has to be informed prior to invasive procedures. The approach in the paper can be extended with such a feature.

An early approach using logic-based reasoning about actions and change for modeling and verification of business processes is presented in [30], based on the ConGolog language. The work is in the line of declarative modeling of processes, while our work is aimed at enhancing BPMN-like models with semantic knowledge and reasoning.

Awad et al. [2] developed a framework for the verification of compliance of a BPMN process to requirements expressed with visual patterns, mapped to temporal logic. The requirements may involve data objects and their states. Other than being based on different inference machinery, our work allows for ontological knowledge to be accommodated in modeling and reasoning.

In [10] Calvanese et al. present an approach where decision models in the DMN standard [33] are integrated with domain knowledge. Representation and reasoning in such integrated models can be expressed in a version of the $\mathcal{ALC}$ description logic with datatypes; this provides, among other things, complexity results for reasoning tasks on decision models, such as analyzing their completeness. The integration of decision models into knowledge representation formalisms, and, in particular, the FO(.) language, is also studied in [13]. The integration of decision models in the approach presented in this paper is a subject for future work.

Colombo Tosatto et al. [12] study the complexity of the problem of business process regulatory compliance, considering achievement and maintenance obligations, showing that verifying partial compliance is an NP-complete problem, and verifying full compliance is a co-NP-complete problem. While in this paper we do not deal with obligations, it has to be noticed that different kinds of obligations could be modeled in our temporal action language by suitably introducing deontic fluents, as done in [25], where a deontic temporal extension of ASP is developed. We observed in Sect. 4 that the complexity of checking the satisfiability of a temporal formula over the BP specification is in NP; this is in agreement with the complexity result for partial compliance in [12].

In [11] Calvanese et al. study plan synthesis for a variant of Knowledge and Action Bases (KAB), a dynamic framework introduced in [6] where states are

DL knowledge bases and an initial ABox evolves over time due to actions which have conditional effects. In particular, [11] focuses on state bounded KABs, for which plan existence is proved to be decidable and shows that, for lightweight DLs, plan synthesis can be compiled into ADL planning.

De Masellis et al. in [14] describe a framework for business process verification combining a control flow model based on Petri Nets with a data model à la Data Centric Dynamic systems. In particular, they adopt the data interaction formalism in [6,11] and prove the decidability of reachability (which in general is undecidable) under three notions of state boundedness. The framework is then encoded in a $\mathcal{C}$-based action language. Finiteness of the domain is guaranteed by the fact that the model is state-bounded. In our approach we can consider the domain to be finite (for each fixed bound in the BMC), by assuming that the data type of objects in the business process is finite. We uniformly model in ASP the business process, the action language (including the constraints extracted from ontological domain knowledge, which is not considered in [14]) and the bounded model checking verification for general formulas, which subsumes reachability analysis.

References

1. Artale, A., Kontchakov, R., Ryzhikov, V., Zakharyaschev, M.: DL-Lite with temporalised concepts, rigid axioms and roles. In: 7th International Symposium on Frontiers of Combining Systems, FroCoS 2009, Trento, Italy, 16–18 September 2009, Proceedings, pp. 133–148 (2009)
2. Awad, A., Weidlich, M., Weske, M.: Visually specifying compliance rules and explaining their violations for business processes. J. Vis. Lang. Comput. **22**(1), 30–55 (2011)
3. Baader, F., Brandt, S., Lutz, C.: Pushing the $\mathcal{EL}$ envelope. In: Kaelbling, L., Saffiotti, A. (eds.) Proceedings of IJCAI 2005, pp. 364–369, Edinburgh, Scotland, UK, August 2005
4. Baader, F., Brandt, S., Lutz, C.: Pushing the $\mathcal{EL}$ envelope. In: LTCS-Report LTCS-05-01. Institute for Theoretical Computer Science, TU Dresden (2005)
5. Baader, F., Calvanese, D., McGuinness, D.L., Nardi, D., Patel-Schneider, P.F. (eds.): The Description Logic Handbook: Theory, Implementation, and Applications. Cambridge University Press, Cambridge (2007)
6. Hariri, B.B., Calvanese, D., Montali, M., De Giacomo, G., De Masellis, R., Felli, P.: Description logic knowledge and action bases. J. Artif. Intell. Res. **46**, 651–686 (2013)
7. Baier, C., Katoen, J.: Principles of Model Checking. MIT Press, Cambridge (2008)
8. Baral, C., Gelfond, M.: Reasoning agents in dynamic domains. In: Minker, J. (ed.) Logic-Based Artificial Intelligence, pp. 257–279 (2000)
9. Biere, A., Cimatti, A., Clarke, E.M., Strichman, O., Zhu, Y.: Bounded model checking. Adv. Comput. **58**, 118–149 (2003)
10. Calvanese, D., Dumas, M., Maggi, F.M., Montali, M.: Semantic DMN: formalizing decision models with domain knowledge. In: Rules and Reasoning - International Joint Conference, RuleML+RR 2017, London, UK, 12–15 July 2017, Proceedings, pp. 70–86 (2017)

11. Calvanese, D., Montali, M., Patrizi, F., Stawowy, M.: Plan synthesis for knowledge and action bases. In: Proceedings of the Twenty-Fifth International Joint Conference on Artificial Intelligence, IJCAI 2016, New York, NY, USA, 9–15 July 2016, pp. 1022–1029 (2016)
12. Colombo Tosatto, S., Governatori, G., Kelsen, P.: Business process regulatory compliance is hard. IEEE Trans. Serv. Comput. **8**(6), 958–970 (2015)
13. Dasseville, I., Janssens, L., Janssens, G., Vanthienen, J., Denecker, M.: Combining DMN and the knowledge base paradigm for flexible decision enactment. In: Supplementary Proceedings of the RuleML 2016 Challenge, Doctoral Consortium and Industry Track Hosted by the 10th International Web Rule Symposium, RuleML 2016 (2016)
14. De Masellis, R., Francescomarino, C.D., Ghidini, C., Montali, M., Tessaris, S.: Add data into business process verification: bridging the gap between theory and practice. In: Proceedings of the Thirty-First AAAI Conference on Artificial Intelligence, 4–9 February 2017, San Francisco, California, USA, pp. 1091–1099 (2017)
15. Dovier, A., Formisano, A., Pontelli, E.: Perspectives on logic-based approaches for reasoning about actions and change. In: Logic Programming, Knowledge Representation, and Nonmonotonic Reasoning - Essays Dedicated to Michael Gelfond on the Occasion of His 65th Birthday, pp. 259–279 (2011)
16. Eiter, T., Faber, W., Leone, N., Pfeifer, G., Polleres, A.: A logic programming approach to knowledge-state planning: semantics and complexity. ACM Trans. Comput. Logic **5**(2), 206–263 (2004)
17. Fahland, D., Favre, C., Koehler, J., Lohmann, N., Völzer, H., Wolf, K.: Analysis on demand: instantaneous soundness checking of industrial business process models. Data Knowl. Eng. **70**(5), 448–466 (2011)
18. Gebser, M., Kaminski, R., Kaufmann, B., Schaub, T.: Answer Set Solving in Practice. Morgan & Claypool Publishers, San Rafael (2012)
19. Gebser, M., Ostrowski, M., Schaub, T.: Constraint answer set solving. In: Hill, P.M., Warren, D.S. (eds.) ICLP 2009. LNCS, vol. 5649, pp. 235–249. Springer, Heidelberg (2009). https://doi.org/10.1007/978-3-642-02846-5_22
20. Gelfond, M., Lifschitz, V.: The stable model semantics for logic programming. In: Logic Programming, Proceedings of the 5th International Conference and Symposium, pp. 1070–1080 (1988)
21. Gelfond, M., Lifschitz, V.: Action languages. Artif. Intell. **2**, 193–210 (1998)
22. Giordano, L., Martelli, A., Spiotta, M., Theseider Dupré, D.: Business process verification with constraint temporal answer set programming. Theory Pract. Logic Prog. **13**, 641–655 (2013)
23. Giordano, L., Martelli, A., Spiotta, M., Theseider Dupré, D.: ASP for reasoning about actions with an EL-bot knowledge base. In: Proceedings of the 31st Italian Conference on Computational Logic, Milano, Italy, 20–22 June 2016, pp. 214–229 (2016)
24. Giordano, L., Martelli, A., Theseider Dupré, D.: Reasoning about actions with temporal answer sets. Theory Pract. Logic Prog. **13**, 201–225 (2013)
25. Giordano, L., Martelli, A., Theseider Dupré, D.: Temporal deontic action logic for the verification of compliance to norms in ASP. In: Proceedings of ICAIL 2013 (2013)
26. Giunchiglia, E., Lifschitz, V.: An action language based on causal explanation: preliminary report. Proc. AAAI/IAAI **1998**, 623–630 (1998)
27. Henriksen, J., Thiagarajan, P.: Dynamic linear time temporal logic. Ann. Pure Appl. logic **96**(1–3), 187–207 (1999)

28. Hoffmann, J., Weber, I., Governatori, G.: On compliance checking for clausal constraints in annotated process models. Inf. Syst. Front. **14**, 155–177 (2009)
29. International Health Terminology Standards Development Organization: SNOMED CT. http://www.ihtsdo.org/snomed-ct/
30. Koubarakis, M., Plexousakis, D.: A formal framework for business process modelling and design. Inf. Syst. **27**(5), 299–319 (2002)
31. Ly, L.T., Rinderle-Ma, S., Göser, K., Dadam, P.: On enabling integrated process compliance with semantic constraints in process management systems - requirements, challenges, solutions. Inf. Syst. Front. **14**(2), 195–219 (2012)
32. zur Muehlen, M., Indulska, M.: Modeling languages for business processes and business rules: a representational analysis. Inf. Syst. **35**(4), 379–390 (2010)
33. Object Management Group: Object Management Group: Decision Model and Notation (DMN) 1.0. http://www.omg.org/spec/DMN/1.0/
34. Roa, H., Indulska, M., Sadiq, S.W.: Effectiveness of domain ontologies to facilitate shared understanding and cross-understanding. In: Proceedings of the International Conference on Information Systems - Exploring the Information Frontier, ICIS 2015, Fort Worth, Texas, USA, 13–16 December 2015
35. Sadiq, S., Governatori, G., Namiri, K.: Modeling control objectives for business process compliance. In: Alonso, G., Dadam, P., Rosemann, M. (eds.) BPM 2007. LNCS, vol. 4714, pp. 149–164. Springer, Heidelberg (2007). https://doi.org/10.1007/978-3-540-75183-0_12
36. The Business Rules Group: Defininig business rules - What are they really? http://www.businessrulesgroup.org/first_paper/BRG-whatisBR_3ed.pdf
37. Wagner, G.: Rule modeling and markup. In: Reasoning Web, First International Summer School 2005, Msida, Malta, 25–29 July 2005, Tutorial Lectures. pp. 251–274 (2005)

Track II: Engineering

Alarm-Based Prescriptive Process Monitoring

Irene Teinemaa[1]([✉]), Niek Tax[2], Massimiliano de Leoni[2], Marlon Dumas[1], and Fabrizio Maria Maggi[1]

[1] University of Tartu, Tartu, Estonia
{irene.teinemaa,marlon.dumas,f.m.maggi}@ut.ee
[2] Eindhoven University of Technology, Eindhoven, The Netherlands
{n.tax,m.d.leoni}@tue.nl

Abstract. Predictive process monitoring is concerned with the analysis of events produced during the execution of a process in order to predict the future state of ongoing cases thereof. Existing techniques in this field are able to predict, at each step of a case, the likelihood that the case will end up in an undesired outcome. These techniques, however, do not take into account what process workers may do with the generated predictions in order to decrease the likelihood of undesired outcomes. This paper proposes a framework for prescriptive process monitoring, which extends predictive process monitoring approaches with the concepts of alarms, interventions, compensations, and mitigation effects. The framework incorporates a parameterized cost model to assess the cost-benefit tradeoffs of applying prescriptive process monitoring in a given setting. The paper also outlines an approach to optimize the generation of alarms given a dataset and a set of cost model parameters. The proposed approach is empirically evaluated using a range of real-life event logs.

1 Introduction

Predictive process monitoring [1,2] is a family of techniques to predict the future state of ongoing cases of a business process based on event logs recording past executions thereof. A predictive process monitoring technique may provide predictions on the remaining execution time of an ongoing case, the next activity to be executed, or the final outcome of the case wrt. a set of possible outcomes. This paper is concerned with the latter type of predictive process monitoring, which we call *outcome-oriented* [3]. For example, in a lead-to-order process, an outcome-oriented predictive process monitoring technique may predict whether a case will end up in a purchase order (desired outcome) or not (undesired outcome).

Existing outcome-oriented predictive process monitoring techniques are able to predict, after each event of a case, the likelihood that the case will end up

Work supported by the European Community's FP7 Framework Program under grant n. 603993 (CORE) and by the Estonian Research Council (IUT20-55).

M. Weske et al. (Eds.): BPM Forum 2018, LNBIP 329, pp. 91–107, 2018.
https://doi.org/10.1007/978-3-319-98651-7_6

in an undesired outcome. These techniques are restricted in scope to predicting. They do not suggest nor prescribe how and when process workers should intervene in order to decrease the likelihood of undesired outcomes.

This paper proposes a framework to extend outcome-oriented predictive process monitoring techniques in order to make them prescriptive. Concretely, the proposed framework extends a given outcome-oriented predictive process monitoring model with a mechanism for generating alarms that lead to interventions, which, in turn, mitigate (or altogether prevent) undesired outcomes. The proposed framework is armed with a parameterized cost model that captures, among others, the tradeoff between the cost of an intervention and the cost of an undesired outcome. Based on this cost model, the paper outlines an approach for return on investment analysis of a prescriptive process monitoring system under a configuration of cost parameters and a predictive model trained on a given dataset. Finally, the paper proposes and empirically evaluates an approach to tune the generation of alarms to minimize the expected cost for a given dataset and set of parameters.

The paper is structured as follows. Section 2 introduces basic concepts and notations. Next, Sect. 3 presents the prescriptive process monitoring framework, Sect. 4 outlines the approach to optimize the alarm generation mechanism, and Sect. 5 reports on our empirical evaluation. Finally, Sect. 6 discusses related work, Sect. 7 delineates the limitations of our framework and consequent future work, and Sect. 8 summarizes the contributions.

2   Background: Events, Traces, and Event Logs

For a given set A, A^* denotes the set of all sequences over A and $\sigma = \langle a_1, a_2, \ldots, a_n \rangle$ a sequence of length n; $\langle \rangle$ is the empty sequence and $\sigma_1 \cdot \sigma_2$ is the concatenation of sequences σ_1 and σ_2. $hd^k(\sigma) = \langle a_1, a_2, \ldots, a_k \rangle$ is the prefix of length k $(0 < k < n)$ of sequence σ. For example, $hd^2(\langle a, b, c, d, e \rangle) = \langle a, b \rangle$.

Let $\mathcal{E}$ be the event universe, i.e., the set of all possible event identifiers, and $\mathcal{T}$ the time domain. We assume that events are characterized by various properties, e.g., an event has a timestamp, corresponds to an activity, is performed by a particular resource, etc. We do not impose a specific set of properties, however, we assume that two of these properties are the timestamp and the activity of an event, i.e., there is a function $\pi_{\mathcal{T}} \in \mathcal{E} \to \mathcal{T}$ that assigns timestamps to events, and a function $\pi_{\mathcal{A}} \in \mathcal{E} \to \mathcal{A}$ that assigns to each event an activity from a finite set of process activities $\mathcal{A}$. An *event log* is a set of events, each linked to one trace and globally unique, i.e., the same event cannot occur twice in a log. A trace in a log represents the execution of one case.

Definition 1 (Trace, Event Log). *A trace is a finite non-empty sequence of events $\sigma \in \mathcal{E}^*$ such that each event appears only once and time is non-decreasing, i.e., for $1 \le i < j \le |\sigma| : \sigma(i) \ne \sigma(j)$ and $\pi_{\mathcal{T}}(\sigma(i)) \le \pi_{\mathcal{T}}(\sigma(j))$. An event log is a set of traces $L \subset \mathcal{E}^*$ such that each event appears at most once in the entire log.*

3   Prescriptive Process Monitoring Framework

In this section, we introduce a cost model for alarm-based prescriptive process monitoring and illustrate this model using three scenarios (Sect. 3.1). We then formalize the concept of alarm system (Sect. 3.2) and discuss conditions under which an alarm system has a positive return on investment (Sect. 3.3).

3.1   Concepts and Cost Model

An alarm-based prescriptive process monitoring system (*alarm system* for short) is a monitoring system that raises an alarm in relation to a running case of a business process, in order to indicate that the case is likely to lead to an undesired outcome. These alarms are handled by process workers who intervene by performing an action (e.g., calling a customer or blocking a credit card) in order to prevent or mitigate the undesired outcome. These actions may have a cost, which we call *cost of intervention*. Instead, if the case ends in a negative outcome, this leads to a cost called *cost of undesired outcome*.

As an example, consider a municipality that needs to collect city taxes. If the inhabitants do not pay their taxes on time, the municipality may run into cash flow issues. Accordingly, in case of an unpaid tax debt (undesired outcome), the municipality may decide to outsource the debt collection to an external collection agency, for which it has to pay a recovery fee. These fees constitute the cost of the undesired outcome. In light of their characteristics and past payment history, certain inhabitants may have a higher risk of missing the payment deadline. Therefore, sending a reminder letter to these high-risk inhabitants may increase the likelihood of receiving the payment on time. However, such an intervention comes with costs related to preparing the letter by an employee (proportional to the employee's hourly salary rate) and the postal costs for sending the letter.

In certain scenarios, the cost of an intervention may increase over time, acknowledging the importance of alarming as early as possible. For instance, in a railway maintenance process, if an alarm about a possible railway disruption is raised early, the problem could be solved with regular maintenance procedures. Conversely, if the alarm is raised when the need for maintenance has become urgent, the maintenance provider could be required to allocate more resources in order to solve the problem on time.

When an alarm is raised, there is a certain probability, but no certainty, that the case will reach an undesired outcome if no intervention is made. If the case does not conclude with an undesired outcome even without interventions, doing the intervention causes unnecessary costs (e.g., a company could lose customers and/or opportunities). The cost related to such unnecessary interventions is referred to as *cost of compensation*. For instance, financial institutions may block credit card payments when they suspect that a card was cloned. However, in some cases, it may happen that the suspicion was unfounded and that the payment was legitimate. If these cases become too frequent, the reputation of the financial institution could be hampered.

The purpose of alarming is to avoid an undesired outcome. However, in several scenarios, it is not possible to fully prevent the cost of the undesired outcome, while the intervention could still help to mitigate it. Based on this rationale, we introduce the concept of *mitigation effectiveness* of an intervention, reflecting the proportion of the cost of an undesired outcome that can be avoided by carrying out the intervention. Oftentimes, the mitigation effectiveness decreases with time, i.e., the earlier the intervention takes place, the higher is the proportion of costs that can be avoided. Consider, for instance, the process of paying unemployment benefits by a social security institution. In this case, the aim of an alarm system could be to notify the institution about citizens who might be receiving unentitled benefits. Since the benefits that have already been issued are unlikely to be recollected, the cost of the undesired outcome cannot be avoided completely. Therefore, it is important to raise the alarm as early as possible, in order to effectively mitigate the cost of the undesired outcome.

An alarm system is intended as a system where cases are continuously monitored. However, since continuous monitoring is impractical, we assume that cases are monitored after each executed event and, therefore, alarms can only be raised after that an event has occurred. In the remainder, each case is identified by a trace σ that is (eventually) recorded in an event log. Definition 2 formalizes the costs defined above. Since costs may depend on the position in the case in which the alarm is raised and/or on other cases being executed, we define the costs as functions over the number of already executed events and over the entire set of cases under execution.

Definition 2 (Alarm-based Cost Model). *An* alarm-based cost model *is a tuple* $(c_{in}, c_{out}, c_{com}, eff)$ *consisting of:*

- *a function* $c_{in} \in \mathbb{N} \times \mathcal{E}^* \times 2^{\mathcal{E}^*} \to \mathbb{R}_0^+$ *modeling the* cost of in*tervention: given a trace* σ *belonging to an event log* L, $c_{in}(k, \sigma, L)$ *indicates the cost of an intervention in* σ *when the intervention takes place after the k-th event;*
- *a function* $c_{out} \in \mathcal{E}^* \times 2^{\mathcal{E}^*} \to \mathbb{R}_0^+$ *modeling the* cost of undesired out*come;*
- *a function* $c_{com} \in \mathcal{E}^* \times 2^{\mathcal{E}^*} \to \mathbb{R}_0^+$ *modeling the* cost of com*pensation;*
- *a function* $eff \in \mathbb{N} \times \mathcal{E}^* \times 2^{\mathcal{E}^*} \to [0,1]$ *modeling the* mitigation eff*ectiveness of an intervention: given a trace* σ *belonging to an event log* L, $eff(k, \sigma, L)$ *indicates the mitigation effectiveness of an intervention in* σ *when the intervention takes place after the k-th event.*

To illustrate the versatility of the above cost model, we discuss three use cases for alarm systems and their corresponding cost model configurations. The first scenario, in Box 1, refers to the provision of unemployment benefits. The cost model for this scenario is based on several discussions with the stakeholders of a real social security institution [4]. The second scenario, in Box 2, refers to the detection of malicious credit card payments in a financial institution. Differently from the previous scenario, in this case, there is a risk of cost of compensation: due to the inconvenience caused by blocking their credit card, customers can switch to competitors. Box 3 refers to the process of predictive maintenance in railway services. This scenario is different from the previous ones because, in this case, the cost of an intervention increases over time.

Box 1 — Scenario "Unemployment Benefits"

In several countries, a social security institution is responsible for the execution of a number of employee-related insurances, such as unemployment benefits. When residents (hereafter customers) become unemployed, they are usually entitled to monthly monetary benefits for a certain period of time. These payments are stopped when the customer reports that he/she has found a new job. Unfortunately, several customers omit to inform the institution about finding a job and, thus, keep receiving benefits they are not entitled to. Those customers are expected to return the amount of benefits that they have received unlawfully. However, in practice, this rarely happens and the overpaid amount is lost to the institution. In light of the above, the social security institution would benefit from an alarm system that would inform about customers who are likely to be receiving unentitled benefits. Let $unt(\sigma)$ denote the amount of unentitled benefits received in a case corresponding to trace σ. Based on discussions with the stakeholders of a real social security institution, we designed the following cost model instantiation for such an alarm system.

Cost of intervention. For the intervention, an employee needs to check if the customer is indeed receiving unentitled benefits and, if so, fill in the forms for stopping the payments. Let S be the employee's average salary rate per time unit; let i_s and i_f denote the positions of the events in σ when the employee started working on the intervention and finished it, respectively. The cost of an intervention can be modeled as: $c_{out}(\sigma, L) = (\pi_\mathcal{T}(\sigma(i_f)) - \pi_\mathcal{T}(\sigma(i_s))) \cdot S$.

Cost of undesired outcome. The total amount of unentitled benefits that the customer would obtain without stopping the payments, i.e., $c_{out}(\sigma, L) = unt(\sigma)$.

Cost of compensation. The social security institution works in a situation of monopoly, which means that the customer cannot be lost because of moving to a competitor, i.e., there is no cost of compensation: $c_{com}(\sigma, L) = 0$.

Mitigation effectiveness. The proportion of unentitled benefits that will not be paid thanks to the intervention: $eff(k, \sigma, L) = \frac{unt(\sigma) - unt(hd^k(\sigma))}{unt(\sigma)}$. Note that this cost function is not employed if there is no undesired outcome (i.e., if $unt(\sigma) = 0$).

Box 2 — Scenario "Financial Institution"

Suppose that the customers of a financial institution use their credit cards to make payments online. Each such transaction is associated with a risk that the transaction is made through a cloned card. In this scenario, an alarm system is intended to determine whether the credit card needs to be blocked due to a high risk of being cloned. However, in case the credit card is not malicious, blocking the card would cause discomfort to the customer who may consequently opt to switch to a different financial institution. Let σ be the trace of credit card transactions for a customer and $value(\sigma)$ the total amount of money related to malicious transactions in σ, the following is a possible cost model instantiation for this scenario.

Cost of intervention. The card is automatically blocked by the system and, therefore, the intervention costs are limited to POST_COST, i.e., to the costs for sending a new credit card to the customer by mail: $c_{in}(k, \sigma, L) = \text{POST\_COST}$.

Cost of undesired outcome. The total amount of money related to malicious transactions that the bank would need to reimburse to the legitimate customer: $c_{out}(\sigma, L) = value(\sigma)$.

Cost of compensation. Denoting the asset value of a customer (consisting of the amount of the investment portfolio, the account balance, etc.) with $asset(\sigma)$ and supposing that a fraction p (i.e., $p \in [0, 1]$) of the customers would switch to a different institution, the cost of compensation can be estimated as the value of the lost asset (the customer), multiplied by p: $c_{com} = p \cdot asset(\sigma)$.

Mitigation effectiveness. The proportion of the total amount of money related to malicious transactions that does not need to be reimbursed by blocking the credit card after that k events have been executed: $eff(k, \sigma, L) = \frac{value(\sigma) - value(hd^k(\sigma))}{value(\sigma)}$.

3.2 Alarm-Based Prescriptive Process Monitoring System

An alarm-based prescriptive process monitoring system is driven by the outcome of the cases. Hereon, the outcome of the cases is represented by a function $out \in \mathcal{E}^* \rightarrow \{\text{true}, \text{false}\}$: given a case identified by a trace σ, if the case has an undesired outcome, $out(\sigma) = \text{true}$; otherwise, $out(\sigma) = \text{false}$. In reality, during the execution of a case, its outcome is not yet known and needs to be estimated

---

Box 3 — Scenario "Railway Maintenance"

In a process for railway maintenance, an alarm should be raised when there is a risk that the railway may break down within a relatively short time range. Railway breakdowns can cause severe disruptions in the train transportation (i.e., trains could be canceled or delayed), thereby causing losses of reimbursing tickets to travelers.

Cost of intervention. The cost of an intervention increases with time because the more urgent the disruption is, the more resources need to be allocated for handling it. We assume that the cost is at its minimum m at the beginning of a trace σ and grows exponentially with time: $c_{in}(k, \sigma, L) = m \cdot \beta \exp(\pi_{\mathcal{T}}(\sigma(k)))$ for some $\beta > 0$.

Cost of undesired outcome. Let P be the average total price of tickets sold per time unit; let $i_d(\sigma)$ and $i_m(\sigma)$ be the positions of the events in σ when the disruption tock place and was resolved, respectively. The cost of the undesired outcome can be calculated as P multiplied by the length of the timeframe when the railway service was disrupted: $c_{out}(\sigma, L) = (\pi_{\mathcal{T}}(\sigma(i_m)) - \pi_{\mathcal{T}}(\sigma(i_d))) \cdot P$.

Cost of compensation. Assuming that performing (unnecessary) maintenance actions does not cause inconveniences to the customers, no cost of compensation is present: $c_{com}(\sigma, L) = 0$.

Mitigation effectiveness. A timely intervention fully avoids the undesired outcome: $eff(k, \sigma, L) = 1$ for any $k \in [1, |\sigma|]$.

---

based on past executions that are recorded in an event log $L \subset \mathcal{E}^*$. The outcome estimator is a function $\widehat{out}_L \in \mathcal{E}^* \to [0, 1]$ predicting the likelihood $\widehat{out}_L(\sigma')$ that the outcome of a case that starts with prefix σ' is undesired. We can define an alarm system as a function that returns true or false depending on whether an alarm is raised based on the predicted outcome or not.

Definition 3 (Alarm-Based Prescriptive Process Monitoring System). *Given an event log $L \subset \mathcal{E}^*$, let $\widehat{out}_L$ be an outcome estimator built from L. An alarm-based prescriptive process monitoring system is a function $alarm_{\widehat{out}_L} \in \mathcal{E}^* \to \{true, false\}$. Given a running case identified by a trace σ and with current prefix σ', $alarm_{\widehat{out}_L}(\sigma)$ returns true, if an alarm is raised based on the predicted outcome $\widehat{out}_L(\sigma')$, or false, otherwise.*

For simplicity, we omit the subscript L from $\widehat{out}_L$ and omit $\widehat{out}_L$ from $alarm_{\widehat{out}_L}$ when it is clear from the context. An alarm system can raise an alarm at most once per case, since we assume that already the first alarm triggers an intervention by the stakeholders.

The purpose of an alarm system is to minimize the cost of executing a case. Table 1 summarizes how the cost of a case is determined based on a cost model (cf. Definition 2), on the case outcome, and on whether an alarm was raised or not.

Definition 4 (Cost of Case Execution). *Let $cm = (c_{in}, c_{out}, c_{com}, eff)$ be an alarm-based cost model. Let $out \in \mathcal{E}^* \to \{true, false\}$ be an outcome function.*

Table 1. Cost of a case σ based on its outcome and whether an alarm was raised

	Undesired outcome	Desired outcome
Alarm raised	$c_{in}(k, \sigma, L) + (1 - eff(k, \sigma, L))c_{out}(\sigma, L)$	$c_{in}(k, \sigma, L) + c_{com}(\sigma, L)$
Alarm not raised	$c_{out}(\sigma, L)$	0

Let $alarm \in \mathcal{E}^* \to \{\textsf{true}, \textsf{false}\}$ be an alarm-based prescriptive process monitoring system. Let $L \subset E^*$ be the entire set of complete (i.e., no more running) cases. Let $\sigma \in L$ be a case. Let $\mathcal{I}(\sigma, alarm)$ be the index of the event in σ when the alarm was raised or zero if no alarm was raised:

$$\mathcal{I}(\sigma, alarm)$$

$$= \begin{cases} 0 & if\ \forall k \in [1, |\sigma|-1].\neg alarm(hd^k(\sigma)), \\ 1 & if\ alarm(hd^1(\sigma)), \\ i \in [2, |\sigma|]\ s.t.\ alarm(hd^i(\sigma)) \wedge & otherwise. \\ \quad \forall k \in [1, i-1].\ \neg alarm(hd^k(\sigma)) \end{cases}$$

The cost of execution of case σ supported by the alarm system is:

$$cost(\sigma, L, cm, alarm)$$

$$= \begin{cases} c_{in}(\mathcal{I}(\sigma, alarm), \sigma, L) \\ \quad + (1 - eff(\mathcal{I}(\sigma, alarm), \sigma, L)) \cdot c_{out}(\sigma, L) & out(\sigma) \wedge \mathcal{I}(\sigma, alarm) > 0, \\ c_{in}(\mathcal{I}(\sigma, alarm), \sigma, L) + c_{com}(\sigma, L) & \neg out(\sigma) \wedge \mathcal{I}(\sigma, alarm) > 0, \\ c_{out}(\sigma, L) & out(\sigma) \wedge \mathcal{I}(\sigma, alarm) = 0, \\ 0 & otherwise. \end{cases}$$

Section 4 illustrates how an alarm-based prescriptive process monitoring system can be designed aiming at the minimization of the case execution costs (according to Definition 4).

3.3   Return on Investment Analysis

In this section, we provide an analysis and guidelines that suggest when it is valuable to invest in developing an alarm system, namely, when the return on investment (ROI) is positive. To this aim, we need to compare the case of a business process execution supported by an alarm system with the *as-is* situation where the business process is executed without this support. For this analysis, we consider a set of cases recorded in an event log L, where no interventions were done, and a cost model $cm = (c_{in}, c_{out}, c_{com}, eff)$.

The *as-is* situation implies that no interventions are done in any of the cases $\sigma \in L$ that lead to an undesired outcome, yielding a cost $c_{out}(\sigma)$. When applied to the entire log L, the cost is $cost_{as\text{-}is}(L) = \sum_{\sigma \in L\ s.t.\ out(\sigma)} c_{out}(\sigma)$. Instead, when a certain system $alarm$ is in effect, the costs are $cost_{alarm}(L) = \sum_{\sigma \in L} cost(\sigma, L, cm, alarm)$ (cf. Definitions 2 and 3). With this setting, the return on investment of the system $alarm$ is $ROI(L, cm, alarm) = cost_{as\text{-}is}(L) - cost_{alarm}(L)$, which must be positive to make deploying the system worthwhile.

The question that remains is: *how does the ROI depend on the cost model and the alarm system?* For the sake of simplicity, we assume, in this analysis, that every component of the cost model is constant. Furthermore, the initial investment costs are not considered because we assume the system to be fully operational already for a sufficiently long time, so that the the initial costs have been amortized. The above assumptions yield the following case cost:

$$cost(\sigma, L, cm, alarm) = \begin{cases} c_{in} + (1 - eff)c_{out} & out(\sigma) \wedge \mathcal{I}(\sigma, alarm) > 0, \\ c_{in} + c_{com} & \neg out(\sigma) \wedge \mathcal{I}(\sigma, alarm) > 0, \\ c_{out} & out(\sigma) \wedge \mathcal{I}(\sigma, alarm) = 0, \\ 0 & \text{otherwise} \end{cases}$$

where c_{in}, c_{out}, c_{com}, and eff are constants. In order for the ROI to be positive, it should be $cost_{as\text{-}is}(L) > cost(\sigma, L, cm, alarm)$, that is:

$$|L_{und}| \cdot c_{out} > |L_{und\&al}|(c_{in} + (1 - eff)c_{out}) + |L_{des\&al}|(c_{in} + c_{com})$$
$$+|L_{und\&nal}| \cdot c_{out}$$

where $L_{und\&al}$, $L_{des\&al}$, $L_{und\&nal}$ respectively consist of the traces in L related to the cases with an <u>und</u>esired outcome that would be <u>al</u>armed, with a <u>des</u>ired outcome that would still be <u>al</u>armed, with an <u>und</u>esired outcome that would <u>n</u>ot be <u>al</u>armed; also, $L_{und} = L_{und\&al} \cup L_{und\&nal}$. After simplification:

$$|L_{und\&al}|(eff c_{out} - c_{in}) > |L_{des\&al}|(c_{in} + c_{com}). \tag{1}$$

Because the right-hand side of Eq. 1 is non-negative, it follows as a corollary that $eff c_{out} > c_{in}$ is a necessary condition for return on investment. In other words, it must be possible to avoid a cost that is higher than the cost of doing the intervention. This provides a validation of our framework: it complies with the *reasonableness condition* in the cost-sensitive learning literature [5], which states that the cost of labeling an example incorrectly should always be greater than the cost of labeling it correctly.

Equation 1 also illustrates that the policy of always alarming does not yield a positive ROI, unless the number of cases with undesired outcome and the cost of the undesired outcome are sufficiently high. When the number of cases with an undesired outcome is small (e.g., the unemployment benefits and the financial institution scenarios described in Boxes 1 and 2) and at the same time the cost of this undesired outcome is small, then the left-hand side of Eq. 1 is negligible, thus leading to condition $c_{in} + c_{com} < 0$, which can never hold.

So far we have assumed, for the sake of simplicity, that costs and mitigation effectiveness are constant, similarly to traditional cost-sensitive learning. However, the novelty of our formulation lays in the fact that costs are functions that depend on the time when an intervention is made. As a result, the reasonableness of the cost matrix would not be fixed, but potentially changes over time. Still, variable costs do not invalidate the ROI analysis. In fact, in order for the ROI to be positive, it is sufficient that the cost model is reasonable for a certain time period; otherwise, the alarm system would never raise alarms because of the cost model. Clearly, the longer the reasonable-cost period is, the higher the ROI.

4   Alarming Mechanisms and Empirical Thresholding

An alarm system needs two components to minimize the costs of future cases: (1) a probabilistic classifier $\widehat{out}_L \in \mathcal{E}^* \to [0,1]$ that estimates the likelihood of an undesired outcome for a partial trace based on some historical observations L, and (2) an alarming mechanism that, for a given incomplete case, decides whether or not to raise an alarm based on the prediction made by $\widehat{out}_L$. We propose to implement the second component using a function $agent \in [0,1] \to$ {true, false} that operates on the estimated likelihood of an undesired outcome, where value **true** represents the decision to raise an alarm. Together, the two components form an *alarm system*, $alarm(hd^k(\sigma)) = agent(\widehat{out}_L(hd^k(\sigma)))$, which makes the decision on whether or not to raise an alarm based on the observed k events of trace σ.

The first component, function $\widehat{out}_L$, can be implemented using any classification algorithm that is naturally probabilistic, i.e., that outputs likelihood scores on a $[0,1]$-interval instead of a binary outcome. Examples of probabilistic classification algorithms include naive Bayes, logistic regression, and random forest. The classifier is trained on historical cases recorded in a log L_{train}.

It is easy to see that the decision on whether or not to raise an alarm should be dependent not only on $\widehat{out}_{L_{train}}(hd^k(\sigma))$, but also on the configuration of c_{in}, c_{out}, c_{com}, and eff. When c_{in} and c_{com} are very low compared to c_{out}, it might be beneficial to use a lower threshold for the estimated likelihood $\widehat{out}_{L_{train}}(hd^k(\sigma))$, while one would want to be more certain that the undesired outcome will happen when c_{in} or c_{com} is high.

We propose to implement the second component, *agent*, as an *alarming threshold*, i.e., a mechanism that alarms when the estimated likelihood of an undesired outcome is at least τ. We define function $alarm_\tau(hd^k(\sigma))$ to be the alarming function that uses the alarming mechanism $agent_\tau(\widehat{out}_{L_{train}}(hd^k(\sigma))) = \widehat{out}_{L_{train}}(hd^k(\sigma)) \geq \tau$. We aim at finding the optimal value $\bar{\tau}$ of the alarming threshold that minimizes the cost on a log L_{thres} consisting of historical observations such that $L_{thres} \cap L_{train} = \emptyset$ with respect to a given likelihood estimator $\widehat{out}_{L_{train}}$ and cost model cm. The total cost of an alarming mechanism $alarm$ on a log L is defined as $cost(L, cm, alarm) = \Sigma_{\sigma \in L} cost(\sigma, L, cm, alarm)$. Using this definition, we define $\bar{\tau} = \arg\min_{\tau \in [0,1]} cost(L_{thres}, cm, alarm_\tau)$. Optimizing a threshold τ on a separate thresholding set is called *empirical thresholding* [6] and the search for the optimal threshold $\bar{\tau}$ wrt. a specified cost model and log L_{thres} can be performed using any hyperparameter optimization technique, such as Tree-structured Parzen Estimator (TPE) optimization [7]. The resulting approach can be considered to be a form of cost-sensitive learning, since the value $\bar{\tau}$ depends on how the cost model cm is specified.

Note that as an alternative to a single global alarming threshold $\bar{\tau}$ it is possible to optimize a separate threshold $\bar{\tau_k}$ for each prefix length k. We experimentally found a single global threshold $\bar{\tau}$ optimized on L_{thres} to outperform separate prefix-length-dependent thresholds $\bar{\tau_k}$ optimized on L_{thres}, therefore we propose to use a single optimized threshold.

After creating the fully functional alarm system by training a classifier on L_{train} and optimizing the alarming threshold on L_{thres} for the given cost model cm, the obtained alarming function $alarm$ can be applied to the continuous stream of events coming from the executions of a business process, thereby reducing the processing costs of the running cases.

5   Evaluation

In this section, we describe the experimental setup for evaluating the proposed framework and the results of the evaluation. We address the following research questions:

RQ1. Can empirical thresholding find thresholds that consistently lead to a reduction in the average processing cost for different cost model configurations?

RQ2. Does the alarm system consistently yield a benefit over different values of the mitigation effectiveness?

RQ3. Does the alarm system consistently yield a benefit over different values of the cost of compensation?

5.1   Approaches and Baselines

We experiment with two different implementations of $\widehat{out}_{L_{train}}$ by using different well-known classification algorithms, namely, random forest (RF) and gradient boosted trees (GBT). Both classification algorithms have shown to be amongst the top performing classification algorithms on a variety of classification tasks [8, 9]. We employ a single classifier approach where the features for a given prefix are obtained using the aggregation encoding [10], which has been shown to perform better than alternative encodings for event logs [3].

We apply the TPE optimization procedure for the alarming mechanism to find the optimal threshold $\bar{\tau}$. We use several fixed thresholds as baselines. First, we compare with the *as-is* situation in which alarms are never raised. Secondly, we compare with the baseline $\tau = 0$, allowing us to compare with the situation where alarms are always raised directly at the start of a case. Finally, we compare with $\tau = 0.5$ enabling the comparison with the cost-insensitive scenario that simply alarms when an undesired outcome is expected. The implementation of the approach and the experimental setup are openly available online.[1]

5.2   Datasets

For each event log, we use all available data attributes as input to the classifier. Additionally, we extract the *event number*, i.e., the index of the event in the given case, the *hour, weekday, month, time since case start*, and *time since last*

---

[1] https://taxxer.github.io/AlarmBasedProcessPrediction/.

event. Infrequent values of categorical attributes (occurring less than 10 times in the log) are replaced with value "other", to avoid exploding the dimensionality. Missing attributes are imputed with the respective most recent (preceding) value of that attribute in the same trace when available, otherwise with zero. Traces are cut before the labeling of the case becomes trivially known and are truncated at the 90th percentile of all case lengths to avoid bias from very long traces. We use the following datasets to evaluate the alarm system:

BPIC2017. This log records execution traces from a loan application process in a Dutch financial institution.[2] The event log was split into two sub-logs, denoted with *bpic2017_refused* and *bpic2017_cancelled*. In the first one, the undesired cases refer to the process executions in which the applicant has refused the final offer(s) by the financial institution and, in the second one, the undesired cases consist of those cases where the financial institution has cancelled the offer(s).

Road traffic fines. This event log originates from the Italian local police.[3] The desired outcome is that a fine is fully paid, while in the undesired cases the fine needs to be sent for credit collection.

Unemployment. This event log corresponds to the *Unemployment Benefits* scenario (Box 1 in Sect. 3.1). The undesired outcome is that a resident will receive more benefits than entitled, causing the need for a reclamation. Privacy constraints prevent us from making this event log publicly available.

Table 2 describes the characteristics of the event logs used. The classes are well balanced in *bpic2017_cancelled* and *traffic_fines*, while the undesired outcome is more rare in case of *unemployment* and *bpic2017_refused*. In *traffic_fines*, the traces are very short, while in the other datasets the traces are longer.

Table 2. Dataset statistics

Dataset name	# traces	Class ratio	Min length	Med length	(Trunc.) max length	# events
bpic2017_refused	31 413	0.12	10	35	60	1 153 398
bpic2017_cancelled	31 413	0.47	10	35	60	1 153 398
traffic_fines	129 615	0.46	2	4	5	445 959
unemployment	34 627	0.2	1	21	79	1 010 450

5.3 Experimental Setup

We apply a temporal split, i.e., we order the cases by their start time and from the first 80% of the cases randomly select 80% (i.e., 64% of the total) for L_{train} and 20% (i.e., 16% of the total) for L_{thres}, and use the remaining 20% as the test

---

[2] https://doi.org/10.4121/uuid:5f3067df-f10b-45da-b98b-86ae4c7a310b.
[3] https://doi.org/10.4121/uuid:270fd440-1057-4fb9-89a9-b699b47990f5.

set L_{test}. The events in cases in L_{train} and L_{thres} that overlap in time with L_{test} are discarded in order to not use any information that would not be available yet in a real setting. We use TPE with 3-fold cross validation on L_{train} to optimize the hyperparameters for RF and GBT. We optimize the alarming threshold $\bar{\tau}$ by building the final classifiers using all the traces in L_{train} and search for $\bar{\tau}$ using L_{thres}.

It is common in cost-sensitive learning to apply calibration techniques to the resulting classifier in order to obtain accurate probability estimates and, therefore, more accurate estimates of the expected cost [11]. However, we found that calibrating the classifier using Platt scaling [12] does not consistently improve the estimated likelihood of undesired outcome on the four event logs, and frequently even leads to less accurate likelihood estimates. Therefore, we decided to skip the calibration step. Moreover, since we use empirical thresholding, it is not necessary that the probabilities are well calibrated and it is sufficient that the likelihoods are reasonably ordered.

Table 3 shows the configurations of the cost model that we explore in the evaluation. To answer RQ1, we vary the ratio between $c_{out}(\sigma, L)$ and $c_{in}(k, \sigma, L)$ (keeping $c_{com}(\sigma, L)$ and $eff(k, \sigma, L)$ unchanged). To answer RQ2, we vary both $eff(k, \sigma, L)$ and the ratio between $c_{out}(\sigma, L)$ and $c_{in}(k, \sigma, L)$. To answer RQ3, we vary two ratios: 1) between $c_{out}(\sigma, L)$ and $c_{in}(k, \sigma, L)$ and 2) between $c_{in}(k, \sigma, L)$ and $c_{com}(\sigma, L)$.

We measure the average processing cost per case in L_{test}, and aim at minimizing this cost. Additionally, we measure the *benefit* of the alarm system, i.e., the reduction in the average processing cost of a case when using the alarm system compared to the average processing cost when not using it.

Table 3. Cost model configurations

$c_{out}(\sigma, L)$	$c_{in}(k, \sigma, L)$	$c_{com}(\sigma, L)$	$eff(k, \sigma, L)$
RQ1 $\{1, 2, 3, 5, 10, 20\}$	1	0	$1 - k/|\sigma|$
RQ2 $\{1, 2, 3, 5, 10, 20\}$	1	0	$\{0, 0.1, 0.2, ..., 1\}$
RQ3 $\{1, 2, 3, 5, 10, 20\}$	1	$\{0, 1/20, 1/10, 1/5, 1/2, 1, 2, 5, 10, 20\}$	$1 - k/|\sigma|$

5.4 Results

Figure 1 shows the average cost per case when increasing the ratio of $c_{out}(\sigma, L)$ and $c_{in}(k, \sigma, L)$ from left to right. We only present the results obtained with GBT as we found it to slightly outperform RF. When the ratio between these two costs is balanced (i.e., 1:1), the minimal cost is obtained by never alarming. This is in agreement with the ROI analysis, where we found $eff c_{out} > c_{in}$ to be a necessary condition for having an advantage from an alarm system.

When $c_{out} \gg c_{in}$ the best strategy is to always alarm. When c_{out} is slightly higher than c_{in} the best strategy is to sometimes alarm based on out. We found that the optimized $\bar{\tau}$ always outperforms the baselines. An exception is ratio 2:1 for *traffic_fines*, where never alarming is slightly better.

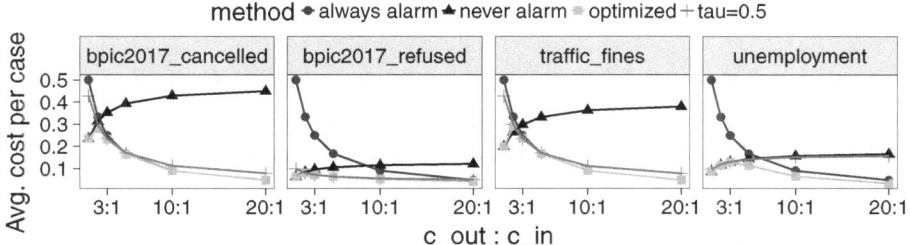

Fig. 1. Cost over different ratios of $c_{out}(\sigma, L)$ and $c_{in}(k, \sigma, L)$ (GBT)

In Fig. 2, the average cost per case is plotted against different (fixed) thresholds. The optimized threshold is marked with a red cross and each line represents one particular cost ratio. We observe that, while the optimized threshold generally obtains minimal costs, there sometimes exist multiple optimal thresholds for a given cost model configuration. For instance, in the case of the 5:1 ratio in *bpic2017_cancelled*, all thresholds between 0 and 0.4 are cost-wise equivalent. We conclude that the empirical thresholding approach consistently finds a threshold that yields the lowest cost in a given event log and cost model configuration (cf. RQ1).

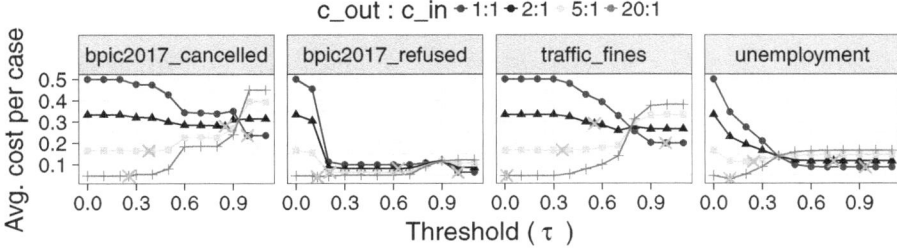

Fig. 2. Cost over different thresholds ($\bar{\tau}$ is marked with a red cross)

Figure 3a shows the benefit of having an alarm system compared to not having it for different (constant) mitigation effectiveness values. As the results are similar for logs with similar class ratios, hereinafter, we only show the results for one log from each of the groups: *bpic2017_cancelled* (balanced classes) and *unemployment* (imbalanced classes). As expected, the benefit increases

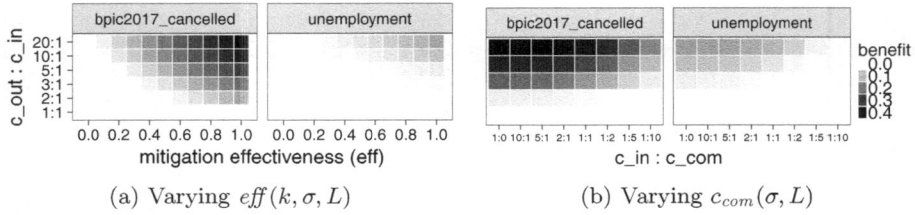

(a) Varying $eff(k, \sigma, L)$ (b) Varying $c_{com}(\sigma, L)$

Fig. 3. Benefit with different cost model configurations

both with higher $eff(k, \sigma, L)$ and with higher $c_{out}(\sigma, L) : c_{in}(k, \sigma, L)$ ratio. For *bpic2017_cancelled*, the alarm system yields a benefit when $c_{out}(\sigma, L) : c_{in}(k, \sigma, L)$ is high and $eff(k, \sigma, L) > 0$. Also, a benefit is always obtained when $eff(k, \sigma, L) > 0.5$ and $c_{out}(\sigma, L) > c_{in}(k, \sigma, L)$. In the case of *unemployment*, the average benefits are smaller, since there are fewer cases with undesired outcome and, therefore, the number of cases where c_{out} can be prevented by alarming is lower. In this case, a benefit is obtained when both $eff(k, \sigma, L)$ and $c_{out}(\sigma, L) : c_{in}(k, \sigma, L)$ are high. We conducted analogous experiments with linear effectiveness decay, varying the maximum possible effectiveness (at the start of the case), which confirmed that the observed patterns remain the same. We have empirically confirmed our theoretical finding (Sect. 3.3) that $eff c_{out} > c_{in}$ is a necessary condition to obtain a benefit from using an alarm system, and have shown that a benefit is in practice also obtained under this condition when an optimized alarming threshold is used (cf. RQ2).

Similarly, the benefit of the alarm system is plotted in Fig. 3b across different ratios of $c_{out}(\sigma, L) : c_{in}(k, \sigma, L)$ and $c_{in}(k, \sigma, L) : c_{com}(\sigma, L)$. We observe that when $c_{com}(\sigma, L)$ is high, the benefit decreases due to false alarms. For *bpic2017_cancelled*, a benefit is obtained almost always, except when $c_{out}(\sigma, L) : c_{in}(k, \sigma, L)$ is low (e.g., 2:1) and $c_{com}(\sigma, L)$ is high (i.e., higher than $c_{in}(k, \sigma, L)$). For *unemployment*, a benefit is obtained with fewer cost model configurations, e.g., when $c_{out}(\sigma, L) : c_{in}(k, \sigma, L) = 5 : 1$ and $c_{com}(\sigma, L)$ is smaller than $c_{in}(k, \sigma, L)$. We conducted analogous experiments with linearly increasing cost of intervention, varying the maximum possible cost (at the end of the case), which confirmed that the patterns described above remain the same. To answer RQ3, we have empirically confirmed that the alarm system achieves a benefit as discussed in Sect. 3.3 in case the cost of the undesired outcome is sufficiently higher than the cost of the intervention and/or the cost of the intervention is sufficiently higher than the cost of compensation.

6   Related Work

The problem of cost-sensitive training of machine learning models has received significant attention. For example, Elkan [5] analyzes the notion of misclassification cost and defines conditions under which a misclassification cost matrix is reasonable. Turney [13] examines a broader range of costs in the context of inductive concept learning. This latter study introduces the notion of cost of intervention, which we include in our proposed cost model. These approaches, however, do not take into account the specific costs that arise in prescriptive process monitoring.

Predictive and prescriptive process monitoring are related to Early Classification of Time Series (ECTS), which aims at classifying a (partial) time series as early as possible, while achieving high classification accuracy [14]. To the best of our knowledge, works [15–17] are the only ECTS methods trying to balance accuracy-related and earliness-related costs. However, these approaches assume that predicting a positive class early has the same effect on the cost function as predicting a negative class early, which is not the case in typical business process monitoring scenarios, where earliness matters only when an undesired outcome is predicted.

Works [18, 19] focus on alarm-based prescriptive process monitoring, but only allow alarms to be raised when a given state of the process is reached. This moment might potentially be late to mitigate the consequences, which would have been possible if the alarm was raised earlier. Furthermore, our approach does not require an explicit modelling of the process states. Last but not least, they rely on a fixed-threshold alarming mechanism provided by process owners, as opposed to our empirical thresholding approach. Gröger et al. [20] is an existing approach to provide recommendations, but it misses the two core elements of our proposed prescriptive process monitoring framework, i.e., cost models and earliness.

7 Limitations and Future Work

While the scenarios discussed in Boxes 1–3 show that the proposed framework is versatile enough to cover a variety of cases, the current version of the framework relies on two main assumptions. First, it assumes that an alarm always triggers an intervention, thus ignoring that a process worker might in some cases decide not to or be unable to intervene. Additionally, the current version of the framework considers each case in isolation, omitting the overall workload of the process workers, which in reality is an important factor for determining the number of alarms that can be acted upon. This limitation can be lifted by, e.g., combining the alarm system with [21], which proposes a recommender system that optimizes suggestions in case of concurrent process executions. A second limitation of the framework is that only one possible type of intervention is envisaged. This assumption can be lifted by extending the framework so that the cost of an intervention can vary depending on the specific action suggested by a recommender system.

Next to these limitations, we acknowledge the importance of further investigation on the applicability of the framework in practice. In particular, in the future, we aim at collaborating with companies and institutions to study whether process stakeholders are able to define the costs in a natural and simple way. Also, we plan to further investigate the consequences of incorrect and/or imprecise instantiations of the cost models. Furthermore, the current evaluation is limited to measuring the benefit of the alarm system in an offline manner, while a more thorough evaluation would consist in deploying the alarming mechanism in a real organization and making an end-to-end comparison of the costs before and after the deployment of the alarm system. However, this is a difficult task for two main reasons. First, companies need to be willing to let the technique really influence the process executions. Second, the end-to-end effectiveness analysis cannot be conducted without coupling the alarm system with a recommender system: if the system raises proper alarms, but inappropriate interventions are taken, the system would still be ineffective. Another avenue for future work is to extend the framework with active learning methods in order to incrementally tune the alarming mechanism based on feedback about the relevance of the alarms and the effectiveness of the interventions.

8   Conclusion

This paper outlined an alarm-based prescriptive process monitoring framework that extends existing predictive process monitoring approaches with the concepts of alarms, interventions, compensations, and mitigation effects. The framework incorporates a cost model to analyze the tradeoffs between the cost of intervention, the benefit of mitigating or preventing undesired outcomes, and the cost of compensating for unnecessary interventions induced by false alarms. The cost model allows one to estimate the benefits of deploying a prescriptive process monitoring system for the purposes of return on investment analysis. Additionally, the framework incorporates a technique to optimize the alarm generation mechanism with respect to a given configuration of the cost model and a given event log. An empirical evaluation on real-life logs showed the benefits of applying this optimization versus a baseline where a fixed likelihood score threshold is used to generate alarms, as considered in previous work in the field.

References

1. Maggi, F.M., Di Francescomarino, C., Dumas, M., Ghidini, C.: Predictive monitoring of business processes. In: Jarke, M., et al. (eds.) CAiSE 2014. LNCS, vol. 8484, pp. 457–472. Springer, Cham (2014). https://doi.org/10.1007/978-3-319-07881-6_31
2. Metzger, A., et al.: Comparing and combining predictive business process monitoring techniques. IEEE Trans. Syst. Man Cybern. Syst. 45(2), 276–290 (2015)
3. Teinemaa, I., Dumas, M., La Rosa, M., Maggi, F.M.: Outcome-oriented predictive process monitoring: review and benchmark. arXiv preprint arXiv:1707.06766 (2017)
4. Dees, M., de Leoni, M., Mannhardt, F.: Enhancing process models to improve business performance: a methodology and case studies. In: Panetto, H., et al. (eds.) OTM 2017. LNCS, vol. 10573, pp. 232–251. Springer, Cham (2017). https://doi.org/10.1007/978-3-319-69462-7_15
5. Elkan, C.: The foundations of cost-sensitive learning. In: Proceedings of IJCAI, vol. 17, pp. 973–978. Lawrence Erlbaum Associates Ltd. (2001)

6. Sheng, V.S., Ling, C.X.: Thresholding for making classifiers cost-sensitive. In: AAAI, pp. 476–481 (2006)
7. Bergstra, J.S., Bardenet, R., Bengio, Y., Kégl, B.: Algorithms for hyper-parameter optimization. In: Proceedings of NIPS, pp. 2546–2554 (2011)
8. Fernández-Delgado, M., Cernadas, E., Barro, S., Amorim, D.: Do we need hundreds of classifiers to solve real world classification problems. JMLR **15**(1), 3133–3181 (2014)
9. Olson, R.S., La Cava, W., Mustahsan, Z., Varik, A., Moore, J.H.: Data-driven advice for applying machine learning to bioinformatics problems. In: Proceedings of Biocomputing. World Scientific (2017)
10. de Leoni, M., van der Aalst, W.M.P., Dees, M.: A general process mining framework for correlating, predicting and clustering dynamic behavior based on event logs. Inf. Syst. **56**, 235–257 (2016)
11. Zadrozny, B., Elkan, C.: Learning and making decisions when costs and probabilities are both unknown. In: Proceedings of KDD, pp. 204–213. ACM (2001)
12. Platt, J., et al.: Probabilistic outputs for support vector machines and comparisons to regularized likelihood methods. Adv. Large Margin Classif. **10**(3), 61–74 (1999)
13. Turney, P.D.: Types of cost in inductive concept learning. In: Proceedings of the Cost-Sensitive Learning Workshop (2002)
14. Xing, Z., Pei, J., Philip, S.Y.: Early classification on time series. KAIS **31**(1), 105–127 (2012)
15. Mori, U., Mendiburu, A., Dasgupta, S., Lozano, J.A.: Early classification of time series by simultaneously optimizing the accuracy and earliness. IEEE Trans. Neural Netw. Learn. Syst. (2017). https://doi.org/10.1109/TNNLS.2017.2764939
16. Dachraoui, A., Bondu, A., Cornuéjols, A.: Early classification of time series as a non myopic sequential decision making problem. In: Appice, A., Rodrigues, P.P., Santos Costa, V., Soares, C., Gama, J., Jorge, A. (eds.) ECML PKDD 2015. LNCS (LNAI), vol. 9284, pp. 433–447. Springer, Cham (2015). https://doi.org/10.1007/978-3-319-23528-8_27
17. Tavenard, R., Malinowski, S.: Cost-aware early classification of time series. In: Frasconi, P., Landwehr, N., Manco, G., Vreeken, J. (eds.) ECML PKDD 2016. LNCS (LNAI), vol. 9851, pp. 632–647. Springer, Cham (2016). https://doi.org/10.1007/978-3-319-46128-1_40
18. Metzger, A., Föcker, F.: Predictive business process monitoring considering reliability estimates. In: Dubois, E., Pohl, K. (eds.) CAiSE 2017. LNCS, vol. 10253, pp. 445–460. Springer, Cham (2017). https://doi.org/10.1007/978-3-319-59536-8_28
19. Di Francescomarino, C., Dumas, M., Maggi, F.M., Teinemaa, I.: Clustering-based predictive process monitoring. IEEE Trans. Serv. Comput. (2017). https://doi.org/10.1109/TSC.2016.2645153
20. Gröger, C., Schwarz, H., Mitschang, B.: Prescriptive analytics for recommendation-based business process optimization. In: Abramowicz, W., Kokkinaki, A. (eds.) BIS 2014. LNBIP, vol. 176, pp. 25–37. Springer, Cham (2014). https://doi.org/10.1007/978-3-319-06695-0_3
21. Conforti, R., de Leoni, M., La Rosa, M., van der Aalst, W.M.P., ter Hofstede, A.H.M.: A recommendation system for predicting risks across multiple business process instances. Decis. Support Syst. **69**, 1–19 (2015)

iProcess: Enabling IoT Platforms in Data-Driven Knowledge-Intensive Processes

Amin Beheshti[1]([✉]), Francesco Schiliro[1,2], Samira Ghodratnama[1],
Farhad Amouzgar[1], Boualem Benatallah[3], Jian Yang[1], Quan Z. Sheng[1],
Fabio Casati[4], and Hamid Reza Motahari-Nezhad[5]

[1] Macquarie University, Sydney, Australia
{amin.beheshti,jian.yang,michael.sheng}@mq.edu.au,
{francesco.schiliro,samira.ghodratnama,farhad.amouzgar}@hdr.mq.edu.au
[2] Australia Federal Police, Canberra, Australia
[3] University of New South Wales, Sydney, Australia
boualem@cse.unsw.edu.au
[4] University of Trento, Trento, Italy
fabio.casati@unitn.it
[5] EY AI Lab, Palo Alto, USA
hamid.motahari@ey.com

Abstract. The Internet of Things (IoT), the network of physical objects augmented with Internet-enabled computing devices to enable those objects sense the real world, has the potential to transform many industries. This includes harnessing real-time intelligence to improve risk-based decision making and supporting adaptive processes from core to edge. For example, modern police investigation processes are often extremely complex, data-driven and knowledge-intensive. In such processes, it is not sufficient to focus on data storage and data analysis; and the knowledge workers (e.g., investigators) will need to collect, understand and relate the big data (scattered across various systems) to process analysis: in order to communicate analysis findings, supporting evidences and to make decisions. In this paper, we present a scalable and extensible IoT-Enabled Process Data Analytics Pipeline (namely *iProcess*) to enable analysts ingest data from IoT devices, extract knowledge from this data and link them to process (execution) data. We introduce the notion of process *Knowledge Lake* and present novel techniques to summarize the linked IoT and process data to construct process *narratives*. This enables us to put the first step towards enabling *storytelling* with process data.

Keywords: Process data science · Process Data Analytics
Data-driven business processes · Knowledge-intensive business processes

© Springer Nature Switzerland AG 2018
M. Weske et al. (Eds.): BPM Forum 2018, LNBIP 329, pp. 108–126, 2018.
https://doi.org/10.1007/978-3-319-98651-7_7

1   Introduction

Information processing using knowledge-, service-, and cloud-based systems has become the foundation of the twenty-first-century life. Recently, the focus of process thinking has shifted towards understanding and analyzing process related data captured in various information systems and services that support processes [2,7,8]. The Internet of Things (IoT), i.e., the network of physical objects augmented with Internet-enabled computing devices to enable those objects sense the real world, has the potential to generate large amount of process related data which can transform many industries. This includes harnessing real-time intelligence to improve risk-based decision making and supporting adaptive processes from core to edge. For example, modern police investigation processes are often extremely complex, data-driven and knowledge-intensive. Considering cases such as Boston bombing (USA), the ingestion, curation and analysis of the big data generated from various IoT devices (CCTVs, Police cars, camera on officers on duty and more) could be vital but is not enough: the big IoT data should be linked to process execution data and also need to be related to process analysis. This will enable organizations to communicate analysis findings, supporting evidences and to make decisions.

Current state-of-the-art in analyzing business processes does not provide sufficient data-driven techniques to relate IoT and process related data to process analysis and to improve risk-based decision making in knowledge intensive processes. To address this challenge, in this paper, we present a scalable and extensible IoT-Enabled Process Data Analytics Pipeline to enable analysts to ingest data from IoT devices, extract knowledge from this data and link them to process (execution) data. We present novel techniques to summarize the linked IoT and process data to construct *process narratives*. Finally, we offer a Machine-Learning-as-a-Service layer to enable process analysts to analyze the narratives and dig for facts in an easy way. We adopt a motivating scenario in policing, where a knowledge worker (e.g., a criminal investigator) in a knowledge intensive process (e.g., criminal investigation) will be augmented by smart devices to collect data on the scene as well as locating IoT devices around the investigation location and communicate with them to understand and analyze evidences in real time. This paper includes offering:

- A scalable and extensible IoT-Enabled Process Data Analytics Pipeline to enable analysts to ingest data from IoT devices, extract knowledge from this data and link them to process (execution) data. We leverage our previous work [3,9] to ingest and organize the big IoT and process data in Data Lakes [3] and to automatically contextualize the raw data in the Data Lake and construct a Knowledge Lake [4].
- A framework and algorithms for *summarizing* the (big) process data and constructing process narrative. We present a set of innovative, fine-grained and intuitive analytical services to discover patterns and related entities, and enrich them with complex data structures (e.g., timeseries, hierarchies and subgraphs) to construct *narratives*.

– A spreadsheet-like dashboard to enable analysts interact with narratives and control their resolution in an easy way. We present a machine-learning-as-a-service framework, which enable analysts dig for facts in an easy way.

The rest of the paper is organized as follows. In Sect. 2 we provide the related work and a motivating scenario. We present the IoT-Enabled Process Data Analytics Pipeline in Sect. 3. We discuss the implementation and the evaluation in Sect. 4 before concluding the paper in Sect. 5.

2   Related Work and Motivating Scenario

2.1   Internet of Things

The Internet of Things (IoT) has the potential to transform many industries and enable them to harness real-time intelligence to improve risk-based decision making and to support adaptive processes from core to edge. In IoT, many of the objects that surround us will be connected, and will be sensing the real world. These objects have the potential to generate large amount of data and meta-data which may contain various facts and evidences. These facts and evidences can help knowledge workers understand knowledge intensive processes and make correct decisions [19]. Many of the work in IoT focus on applications such as smart and connected communities [22], industries (e.g., agriculture, food processing, environmental monitoring, automotive, telecommunications, and health) [15], and security and privacy [1]. Mobile crowdsensing and cyber-physical cloud computing presented as two most important IoT technologies in promoting Smart and Connected Communities [22]. Management of IoT data is an important issue in rapidly changing organizations. A set of recent work has been focusing on ingesting the large amount of data generated from IoT devices and store and organize them in big data platforms. For example, Hortonworks DataFlow (hortonworks.com) provides an end-to-end platform that collects and organizes the IoT data in the cloud. Other approaches include Teradata (teradata.com/) and Oracle BigData (oracle.com/bigdata) focus on data management and analytics, and do not related the data to process analysis.

Enabling IoT data in business process analytics, as presented in this paper, is a novel approach to enhance data-driven techniques for improving risk-based decision making in knowledge intensive processes. The novel notions of Knowledge Lake and narrative, presented in this paper, will enable us to put the first step towards enabling *storytelling* with process data. This will enable analysts to ingest data from IoT devices, extract knowledge from this data and related the data to process analysis.

2.2   Data-Driven Processes

The problem of understanding the behavior of information systems as well as the processes and services they support has become a priority in medium and large enterprises. This is demonstrated by the proliferation of tools for the analysis of process executions, system interactions, and system dependencies, and by recent research work in process data warehousing, discovery and mining [24]. Accordingly, identifying business needs and determining solutions to business problems requires the analysis of business process data which in turn will help in discovering useful information and supporting decision making for enterprises. The state-of-the-art in process data analytics focus on various topics such as Warehousing Business Process Data [14], Data Services and DataSpaces [13], Supporting Big Data Analytics Over Process Execution Data [5], Process Spaces [18], Process Mining [24] and Analyzing Cross-cutting Aspects (e.g., provenance) in Processes' Data [6]. In our recent book [8], we provided a complete state-of-the-art in the area of business process management in general and process data analytics in particular. This book provides defrayals on: (i) technologies, applications and practices used to provide process analytics from querying to analyzing process data; (ii) a wide spectrum of business process paradigms that have been presented in the literature from structured to unstructured processes; (iii) the state-of-the-art technologies and the concepts, abstractions and methods in structured and unstructured BPM including activity-based, rule-based, artifact-based, and case-based processes; and (iv) the emerging trend in the business process management area such as: process spaces, big-data for processes, crowdsourcing, social BPM, and process management on the cloud.

Summarization techniques presented in this paper, is a novel approach to enable analysts to understand and relate the big IoT and process data to process analysis in order to communicate analysis findings and supporting evidences in an easy way. The proposed approach will enhance data-driven techniques for improving risk-based decision making in knowledge intensive processes.

2.3   Knowledge-Intensive Processes

Case-managed processes are primarily referred to as semistructured processes, since they often require the ongoing intervention of skilled and knowledgeable workers. Such Knowledge-Intensive Processes, involve operations that heavily reliant on professional knowledge. For these reasons, it is considered that human knowledge workers are responsible to drive the process, which cannot otherwise be automated as in workflow systems [8]. Knowledge-intensive processes almost always involve the collection and presentation of a diverse set of artifacts and capturing the human activities around artifacts. This, emphasizes the artifact-centric nature of such processes. Many approaches [11,16,23] used business artifacts that combine data and process in a holistic manner and as the basic building block. Some of these works [16] used a variant of finite state machines to specify lifecycles. Some theoretical works [11] explored declarative approaches to specifying the artifact lifecycles following an event oriented style. Another line of work in this category, focused on querying artifact-centric processes [17].

Another related line of work is artifact-centric workflows [11] where the process model is defined in terms of the lifecycle of the documents. Some other works [20,21], focused on modeling and querying techniques for knowledge-intensive tasks. Some of existing approaches [20] for modeling ad-hoc processes focused on supporting ad-hoc workflows through user guidance. Some other approaches [21] focused on intelligent user assistance to guide end users during ad-hoc process execution by giving recommendations on possible next steps. Another line of work [6], considers entities (e.g., actors, activities and artifacts) as first class citizens and focuses on the evolution of business artifacts over time. Unlike these approaches, in *iProcess*, we not only consider artifacts as first class citizens, but we take the information-items (e.g., named entities, keywords, etc.) extracted from the content of the artifacts into account.

2.4　Motivating Scenario: Missing People

As the motivating scenario, we focus on the investigation processes around *Missing Persons*. Between 2008 and 2015 over 305,000 people were reported missing in Australia (aic.gov.au/), an average of 38,159 reports each year. In USA (nij.gov/), on any given day, there are as many as 100,000 active missing person's cases. The first few hours following a person's disappearance are the most crucial. The sooner police is able to put together the sequence of events and actions right before the disappearance of the person, the higher the chance of finding the person. This entails gathering information about the person including physical appearance, and activities on social media in the physical/social environments of the person, person's activity data such as phone calls and emails, and information on the person detected by sensors (e.g. CCTVs).

The investigation process is a data-driven, knowledge-intensive and collaborative process. The information associated with an investigation (case process) are usually complex, entailing the collection and presentation of many different types of documents and records. It is also common that separate investigations may impact other investigation processes, and the more evidences (knowledge and facts extracted from the data in the data lake [3]) collected the better related cases can be linked explicitly. Although law enforcement agencies use data analysis, crime prevention, surveillance, communication, and data sharing technologies to improve their operations and performance, in sophisticated and data intensive cases such as missing persons there still remain many challenges. For example, fast and accurate information collection and analysis is vital in law enforcement applications [10,12]. From the policymakers' perspective, this trend calls for the adoption of innovations and technologically advanced business processes that can help law enforcers detect and prevent criminal acts. Enabling IoT data in law enforcement processes will help investigators to access to a potential pool of data evidences. Then, the challenge would be to prepare the big process data for analytics, summarizing the big process data, constructing narratives and enable analysts to link narratives and dig for facts in an easy way.

In this paper, we aim to address this challenge by augmenting police officers with Internet-enabled smart devices (e.g., phones/watches) to assist them in the process of collecting evidences, access to location-based services to identify and locate resources (CCTVs, camera on officers on duty, police cars, drones and more), organize all these islands of data in a Knowledge Lake [4] and feed them into a scalable and extensible IoT-Enabled Process Data Analytics Pipeline.

3   iProcess: IoT-Enabled Process Data Analytics Pipeline

Figure 1 illustrates the IoT-Enabled Process Data Analytics Pipeline framework. In the following we explain the main phases of the iProcess pipeline.

3.1   Process Data-Lake

In order to understand data-driven knowledge-intensive processes, one may need to perform considerable analytics over large hybrid collections of heterogeneous and partially unstructured data that is captured from private (personal/business), social and open data. Enabling IoT data in such processes will maximize the value of data-in-motion and will require dealing with big data organization challenges such as wide physical distribution, diversity of formats, non-standard data models, independently-managed and heterogeneous semantics. In such an environment, analysts may need to deal with a collection of datasets, from relational to NoSQL, that holds a vast amount of data gathered from various data islands, i.e., Data Lake. To address this challenge, we leverage our previous work [3], CoreDB: a Data Lake as a Service, to identify (IoT, Private, Social and Open) data sources and ingest the big process data in the Data Lake. CoreDB manages multiple database technologies (from relational to NoSQL), offers a built-in design for security and tracing, and provides a single REST API to organize, index and query the data and metadata in the Data Lake.

3.2   Process Knowledge-Lake

The rationale behind a Data Lake is to store raw data and let the data analyst decide how to cook/curate them later. We introduce the notion of Knowledge Lake [4], i.e., a contextualized Data Lake, to provide the foundation for big data analytics by automatically curating the raw data in the Data Lake and to prepare them for deriving insights. To achieve this goal, we leverage our previous work [4], to transform raw data (unstructured, semi-structured and structured data sources) into a contextualized data and knowledge that is maintained and made available for use by end-users and applications. The Data Curation APIs [9] in the Knowledge Lake provide curation tasks such as extraction, linking, summarization, annotation, enrichment, classification and more. This will enable us to add features - such as extracting keyword, part of speech, and named entities such as Persons, Locations and Organizations; providing synonyms and stems for extracted information items leveraging lexical knowledge bases for the English language such as WordNet; linking extracted entities to external knowledge bases such as Google Knowledge Graph and Wikidata; discovering similarity among the extracted information items; classifying, indexing, sorting and categorizing data - into the data and knowledge persisted in the Knowledge Lake.

This will enable us, for example, to extract and link information about the missing person from various data islands in the data lake such as the IoT, social and news data sources and to relate them to missing person case. The goal of this

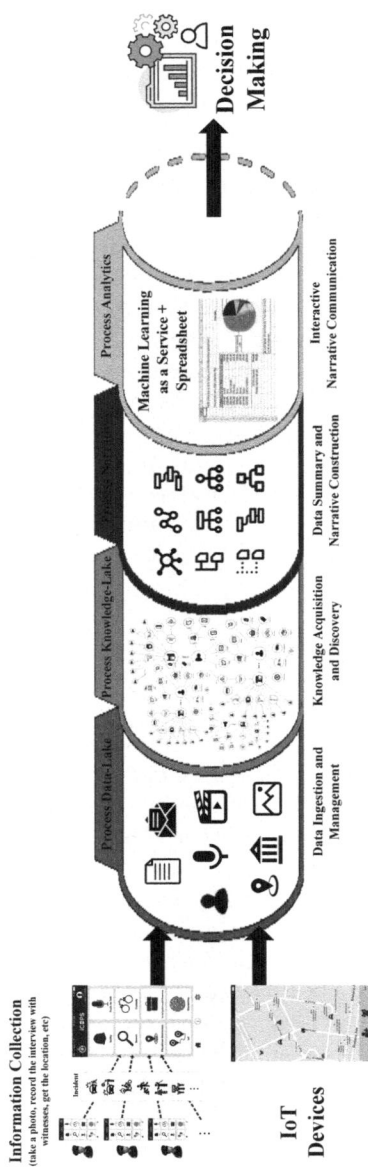

Fig. 1. IoT-Enabled process data analytics pipeline.

phase is to contextualize the Data Lake and turn it into a Process Knowledge-Lake which contains: (i) a set of facts, information, and insights extracted from the raw data; (ii) process event data i.e., observed behavior; and (iii) process models, e.g., manually or automatically discovered. All these three main components will enable the process analysts to relate data to process analysis. To achieve this goal, we present a graph model to define the entities (process data, instances and models) and the relationships among them.

Definition 1 (Process Knowledge Graph). Let $G = (V, E)$ be an Entity-Relationship (ER) attributed graph where V is a set of nodes with $|V| = n$, and $E \subseteq (V \times V)$ is a set of ordered pairs called edges. Let $H = (V, E)$ be a RDF graph where V is a set of nodes with $|V| = n$, and $E \subseteq (V \times V)$ is a set of ordered pairs called edges. An ER graph $G = (V_G, E_G)$ with n entities is defined as $G \subseteq H$, $V_G = V$ and $E_G \subseteq E$ such that G is a directed graph with no directed cycles. We define a resource in an ER graph recursively as follows: (i) The sets V_G and E_G are resources; (ii) $\in$ is a resource; and (iii) The set of ER graphs are closed under intersection, union and set difference: let G_1 and G_2 be two ER graphs, then $G_1 \cup G_2$, $G_1 \cap G_2$, and $G_1 - G_2$ are resources.

Definition 2 (Entity). An entity E is represented as a data object that exists separately and has a unique identity. Entities are described by a set of *attributes* but may not conform to an entity type. Entities can be complex such as Process Model, Process Instance and a (IoT, Social or private) Data Source. One way would be to define "stream events" meaning events that are tied to a specific timestamp or sequence number, and associated to a specific IoT device. Entities can be also simple such as *artifacts* (e.g., structured such as customer record or unstructured such as an email), actors and activities. Entities can be atomic *information items* such as a keyword, phrase, topic and named entity (e.g., people, location, organization) extracted from unstructured artifacts such as emails, images (extracted from IoT devices) or social items (such as a Tweet in Twitter). This entity model offers flexibility when types are unknown and takes advantage of structure when types are known. Entities can be of type stream, such as 'stream events' meaning events that are tied to a specific timestamp or sequence number, and associated to a specific IoT device.

Definition 3 (Relationship). A *relationship* is a directed link between a pair of entities, which is associated with a predicate defined on the attributes of entities that characterizes the relationship. Relationships can be described by a set of *attributes* but may not conform to a relationship type. Relationships can be [2]: Time-based, Content-based and Activity-based. We define the following *explicit* relationships:

– $Process \xrightarrow{\text{(Instance-of)}} Model$: express that a process is an instance of a process model.
– $Process \xrightarrow{\text{(Used)}} Artifact$: express that a process used an artifact during its execution.

- $Artifact \xrightarrow{(Generated\text{-}by)} Process$: express that an artifact was generated by a process.
- $Process \xrightarrow{(Controlled\text{-}by\ (R))} Actor$: express that a process was controlled by an actor. Given that a process may have been controlled by several actors, it is important to identify the roles of actors.
- $Process_1 \xrightarrow{(Triggered\text{-}by)} Process_2$: express a process oriented view where a process triggered another process.
- $Artifact \xrightarrow{(Organized\text{-}in)} Data - Island$: express that an artifact (e.g., an email in a private dataset or an image extracted from a CCTV camera) is organized in a Data Island (i.e., a Data source in the Data Lake).
- $Information - Item \xrightarrow{(Extracted\text{-}from)} Artifact$: express that an information item (e.g., a topic extracted from a Tweet or a named entity such as a person, extracted from an Image) is extracted from an artifact (e.g., an email or an image, extracted from a CCTV camera, in a private data source).
- $Information - Item_1 \xrightarrow{(Similar\text{-}to)} Information - Item_2$: express that an information item (e.g., a person named entity extracted from an Image) is similar to another information item (e.g., a person named entity extracted from an email or a Tweet in Twitter (twitter.com)

Notice that 'Process' refers to a process instance and 'Model' refers to a process model. A *Process Instance or Case*, is a triple $C = (P_F, N_{start}, N_{end})$, where P_F is a path in which the nodes in P are of type 'event', grouped using the function F (e.g. a function can be a 'Correlation Condition'), and are in chronological order. A *Process Model*, allows the generation of all valid (acceptable) case C of a process, e.g. implemented by service or a set of services [2]. Various process mining algorithms and tools (e.g., PROM), include our previous work [18], can be used to automatically extract the first type of relationship. Process instances and services can be instrumented to automatically construct the other type of relationship.

3.3   Process Narratives

In this phase, we present an *OLAP* [5] *style process data summarization* technique as an alternative to querying and analysis techniques. This approach will isolate the process analyst from the process of explicitly analyzing different dimensions such as time, location, activity, actor and more. Instead, the system will be able to use interactive (artifacts, actors, events, tasks, time, location, etc.) summary generation to select and sequence narratives dynamically. This novel summarization method will enable process analysts to choose one or more dimensions (i.e., attributes and relationships), based on their specific goal, and interact with small and informative summaries. This will enable the process analysts to analyze the process from various dimensions. Figure 2(B) illustrates a sample OLAP dimension.

In OLAP [5], cubes are defined as set of partitions, organized to provide a multi-dimensional and multi-level view, where partitions considered as the unit

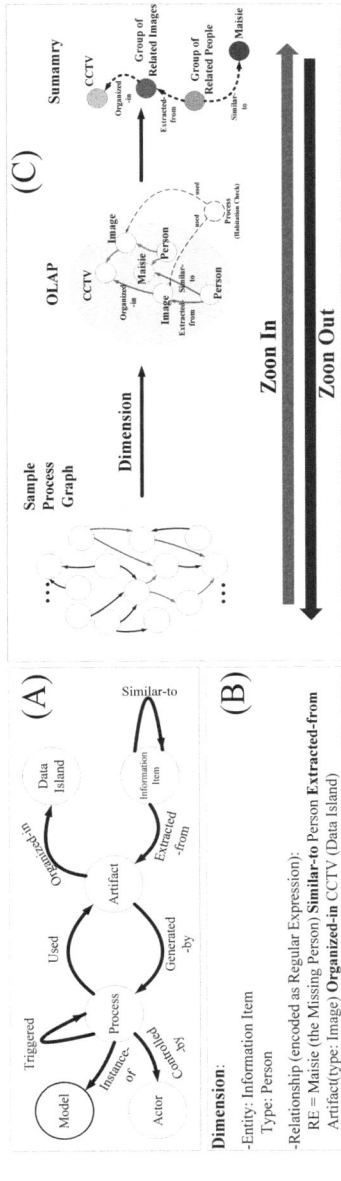

Fig. 2. Process Knowledge Graph schema (A), a sample OLAP Dimension (B) and an interactive graph summary (C).

of granularity. Dimensions defined as perspectives used for looking at the data within constructed partitions. In police investigation scenarios, such as Boston bombing, process cubes can enable effective analysis of the Process Knowledge Graph from different perspectives and with multiple granularities. For example, by aggregating and relating all evidences from the person of interest, location of the incident and more. Following, we define a process cube.

Definition 4 (Process Cube). A process cube defined to extend decision support on multidimensional networks, e.g., process graphs, considering both data objects and the relationships among them. We reuse and extend the definition for graph-cube proposed in our previous work [5]. In particular, given a multidimensional network N, the graph cube is obtained by restructuring N in all possible aggregations of set of node/edge attributes A, where for each aggregation A' of A, the measure is an aggregate network G' w.r.t. A'. We define possible aggregations upon multidimensional networks using Regular Expressions. In particular, $Q = \{q_1, q_2, ..., q_n\}$ is a set of n process cubes, where each q_i is a process cube, a placeholder for set of related entities and/or relationships among them, and can be encoded using regular expressions. In this context, each process cube q_i can extensively support multiple information needs with the graph data model (e.g., Definition 1) and one algorithm (regular language reachability). The set of related process cubes Q is designed to be customizable by local domain experts (who have the most accurate knowledge about their requirement) to codify their knowledge into regular expressions. These expressions can describe paths through the nodes and edges in the attributed graph: Q can be constructed once and can be reused for other processes. The key data structure behind the process cube is the Process Knowledge Graph, i.e., a graph of typed nodes, which represent process related entities (such as process instances, models, artifacts, actors, data sources, and information items), and typed edges, which label the relationships of the nodes to one another, illustrated in Fig. 2(A). We leveraged the graph mining algorithms in our previous work [5] to walk the graph from one set of interesting entities to another via the relationship edges and discover which entities are ultimately transitively connected to each other, and group them in folder nodes (set of related entities) and path nodes (set of related patterns). We use correlation-conditions [18] to partition the Process Knowledge Graph based on set of dimensions coming from the attributes of node entities. We use a path-condition [5] as a binary predicate defined on the attributes of a path that allows to identify whether two or more entities are related through that path.

Definition 5 (Dimensions). Each process cube q_i has a set of dimensions $D = \{d_1, d_2, ..., d_n\}$, where each d_i is a dimension name. Each dimension d_i is represented by a set of elements (E) where elements are the nodes and edges of the Process Knowledge Graph. In particular, $E = \{e_1, e_2, ..., e_m\}$ is a set of m elements, where each e_i is an element name. Each element e_i is represented by a set of attributes (A), where $A = \{a_1, a_2, ..., a_p\}$ is a set of p attributes for element e_i, and each a_i is an attribute name. A dimension d_i can be considered as a given query that require grouping graph entities in a certain way. Correlation-conditions and path-conditions can be used to define such queries.

A dimension uniquely identifies a subgraph in the Process Knowledge Graph, which we call a *Summary*. Now, we introduce the new notion of Narrative.

Definition 6 (Narrative). A narrative $N = \{S, R\}$, is a set summaries $S = \{s_1, s_2, ..., s_n\}$ and a set of relationships $R = \{r_1, r_2, ..., r_m\}$ among them, where s_i is a summary name and r_j is a relationship of type 'part-of' between two summaries. This type of relationship enables the zoom-in and zoom-out operations (see Fig. 2(C)) to link different pieces of a story and enable the analyst to interact with narratives. Each summary $S = \{Dimension, View - Type, Provenance\}$, identified by a unique dimension D, relates to a view type VT (e.g., process, actor or data view) and assigned to a Provenance code snippet P to document the evolution of the summary over time (more nodes and relationships can be added to the Process Knowledge Graph over time). We leverage our work [6] to document the evolution of summaries over time.

The formalism of the summary S will enable to consider different dimensions and views of a narrative, including the event structure (narratives are about something happening), the purpose of a narrative (narratives about actors and artifacts), and the role of the listener (narratives are subjective and depend on the perspective of the process analyst). Also, it considers the importance of time and provenance as narratives may have different meanings over time. We develop a scalable summary generation algorithm and support three types of summaries. Figure 3 illustrates the scalable summary generation process. Following we introduce these summaries:

- Entity Summaries: We use correlation conditions to summarize the Process Knowledge Graph based on set of dimensions coming from the attributes of node entities. In particular, a correlation condition is a binary predicate defined on the attributes of attributed nodes in the graph that allows to identify whether two or more nodes are potentially related. Algorithm 1 in Fig. 3, will generate all possible entity summaries. For example, one possible summary may include all related images captured in the same location. Another summary may include all related images captured in the same timestamp.
- Relationship Summaries: We use correlation conditions to summarize the Process Knowledge Graph based on set of dimensions coming from the attributes of attributed edges. Algorithm 2 in Fig. 3, will generate all possible relationship summaries. For example, one possible summary may include all related relationships typed controlled-by and have the following attributes "Controlled-by (role = 'Investigator'; time = 'τ_1'; location = '255.255.255.0')". In the relationship summaries, we also store the nodes from and to the relationship, e.g., in this example the process instance and the actor.
- Path Summaries: We use path conditions to summarize the Process Knowledge Graph based on set of dimensions coming from the attributes of nodes

and edges in a path, where a path is a transitive relationship between two entities showing a sequence of edges from the start entity to the end. In particular, a path condition defined on the attributes of nodes and edges that allows to identify whether two or more entities (in a given Process Knowledge Graph) are potentially related through that path. Algorithm 3 in Fig. 3, will generate all possible path summaries. For example, one possible relationship summary includes all related images captured in the same location and contain the same information item, e.g., the missing person. Another relationship summary includes all related Tweets or emails sent on timestamp τ_1 and include the keyword Maisie (the missing person).

3.4   Process Analytics

In this phase, we present a spreadsheet like interface on top of the scalable summary generation framework. The goal is to enable analysts to interact with the narratives and control the resolutions of summaries. A narrative N can be analyzed using three operations: (i) roll-up: to aggregate summaries by moving up along one or more dimensions, and to provide a smaller summary with less details. (ii) drill-down: to disaggregate summaries by moving down dimensions; and to provide a larger summary with more details; (iii) slice-and-dice: to perform selection and projection on snapshots. To achieve this goal, we use the notion of spreadsheets and organize all the possible summaries in the rows and columns of a grid. Each tab in the spreadsheet defines a summary type (e.g., entity, relationship or path summary), the rows in a tab are mapped to the dimensions (e.g., Attributes of an entity), and the columns in a tab are mapped to various data islands in the Data Lake. Each cell will contain a specific summary.

We make a set of machine learning algorithms available as a service and to enable the analysts to manipulate and use the summaries in spreadsheets to support: (i) roll-up: the roll-up operation performs aggregation on a spreadsheet tab, either by climbing up a concept hierarchy (i.e., rows and columns which represent the dimensions and data islands accordingly) or by climbing down a concept hierarchy, i.e., dimension reduction; (ii) drill-down: the drill-down operation is the reverse of roll up. It navigates from less detailed summaries to more detailed summaries. It can be realized by either stepping down a concept hierarchy or introducing additional dimensions. For example, in Fig. 4, applying the drill-down operation on the cell intersecting time (dimension) and CCTV1 (data source) will provide a more detailed summary, grouping all the items over different points in time. As another example, applying the drill-down operation on the cell intersecting country (dimension) and Twitter (data source) will provide a more informative summary, grouping all the tweets, twitted in different counties; and (iii) slice-and-dice: the slice operation performs a selection on one dimension of the given tab, thus resulting in a sub-tab. The dice operation defines a sub-tab by performing a selection on two or more dimensions. This will enable analyst, for example to see Tweets coming from 2 dimensions such as time and

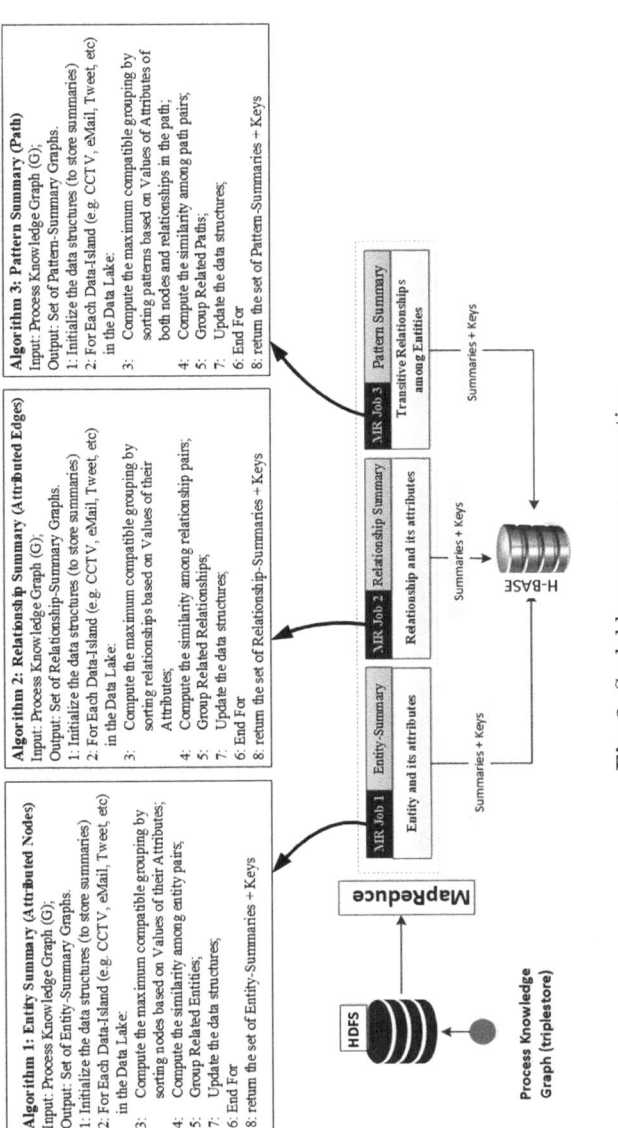

Fig. 3. Scalable summary generation.

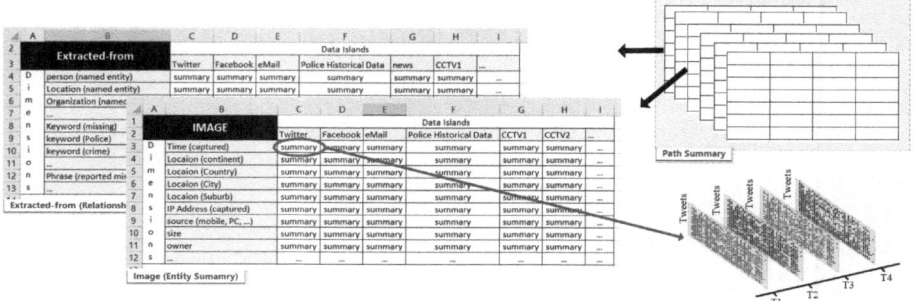

Fig. 4. Presenting a spreadsheet like interface on top of the scalable summary generation framework.

location. The slice-and-dice operation can be simply seen as a regular expression which groups together different entity and/or relationship summaries (presented in the spreadsheet tabs) and weaves them together to construct path summaries, illustrated in Fig. 4.

4   Implementation and Evaluation

We focus on the motivating scenario, to assist knowledge workers in the domain of law enforcement collect information from the investigation scene as well as the IoT-enabled devices of interest in an easy way and on a mobile device. The goal here is to contribute to research and thinking towards making the police officers more effective and efficient at the front-line, while augmenting their knowledge and decision management processes through Information and Communication Technology. We develop ingestion services to extract the raw data from IoT devices such as CCTVs, location sensors in police cars and smart watches (to detect the location of people on duty) and police drones. These services will persist the data in the data lake. Next and inspired by Google Knowledge Graph (developers.google.com/knowledge-graph/), we focused on constructing a policing process knowledge graph: an IoT infrastructure that can collaborate with internet-enabled devices to collect data, understand the events and facts and assist law enforcement agencies in analyzing and understanding the situation and choose the best next step in their processes. There are many systems that can be used at this level including our previous work (Curation APIs) [9], Google Cloud Platform (cloud.google.com/), and Microsoft Computer Vision API (azure.microsoft.com/) to extract information items from artifacts (such as emails, images, social items).

We have identified many useful machine learning algorithms and wrapped them as services to enable us to summarize the constructed knowledge graph, and to extract complex data structures such as timeseries, hierarchies, patterns and subgraphs and link them to entities such as business artifacts, actors, and activities. Figure 5, illustrates the taxonomy of these services. We use a

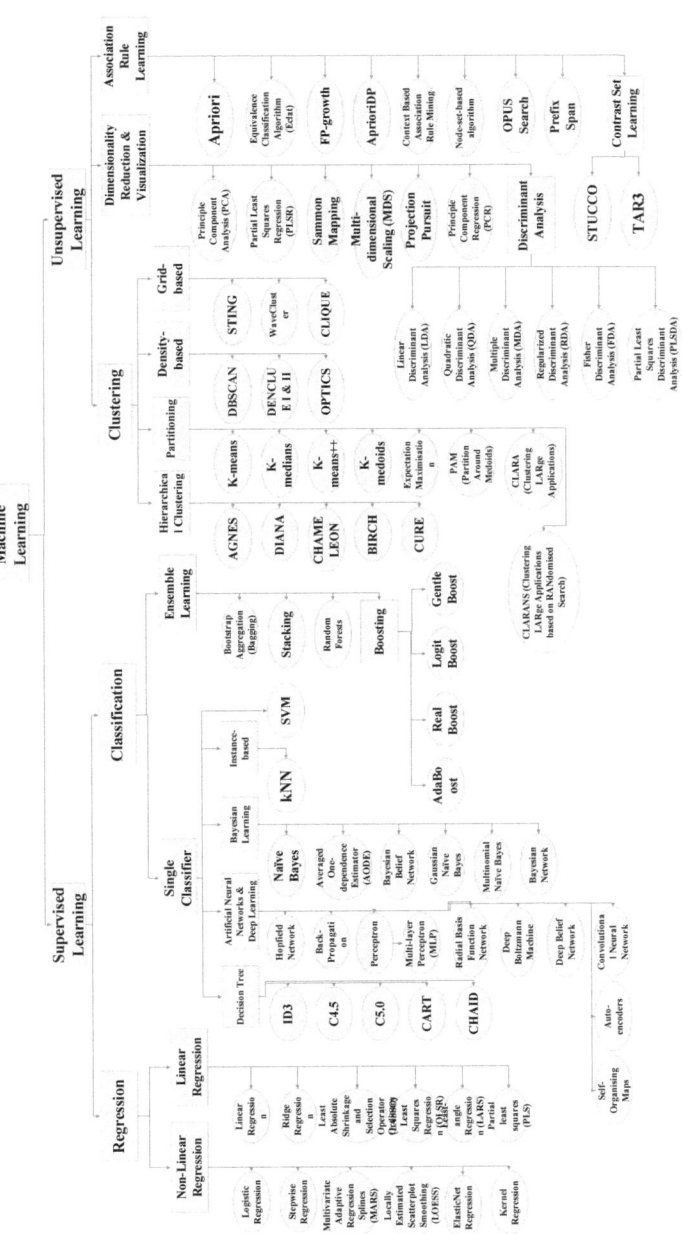

Fig. 5. The Taxonomy of the Machine Learning algorithms used as a Service to enable the knowledge workers interact with the summaries in an easy way.

spreadsheet-like dashboard that enables the knowledge workers interact with the summaries in an easy way. The dashboard enables monitoring the entities (e.g., IoT devices, people, and locations) and dig for the facts (e.g., suspects, evidences and events) in an easy way. A set of services has been developed to link the dashboard to the knowledge graph and the data summaries. A **demonstration** of the prototype can be found in: https://github.com/unsw-cse-soc/CoreKG.

The evaluation of accuracy and performance of the Data Lake and knowledge extraction services demonstrated in [3,9]. Figure 6 shows the performance of our access structure as a function of available memory for entity/relationship and path summaries. These summaries have been generated from a Tweet dataset having over 15 million tweets, persisted and indexed in the MongoDB (mongodb.com) database in our Data Lake. For the path summaries, we have limited the dept of the path to have maximum of three transitive relationship between the starting and ending nodes. The experiment were performed on Amazon EC2 platform using instances running Ubuntu Server 14.04. The memory size is expressed as a percentage of the size required to fit the largest partition of data in the hash access structure in physical memory. For efficient access to single cells (i.e. a summary) we built a partition level hash access structure where the partitions will be kept in memory and the operations will evaluated for one partition at a time. If a summary does not fit in memory we incur an I/O if a referenced cell is not cached. In the case of entity/relationship summary Fig. 6(A), this occurs when the available memory is around 40% of the largest summary, and for the path summary Fig. 6(B) this occurs when the available memory is around 30% of the largest summary.

As future work, we will evaluate the usability of the approach regarding the intended application audience, i.e., the police and expert users.

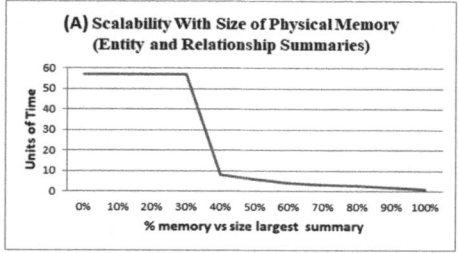 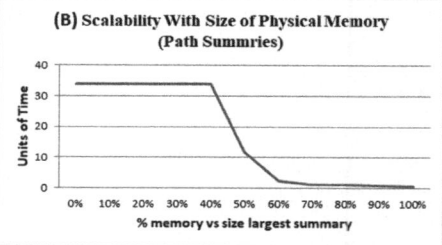

Fig. 6. Scalability with size of physical memory for entity and relationship summaries (A) and scalability with size of physical memory for path summaries (B).

5 Conclusion and Future Work

The large amount of raw data generated by IoT-enabled devices provide real-time intelligence to organizations which can enhance knowledge intensive processes.

For example, one of the interventions that have emerged as a potential solution to the challenges facing law enforcement officers is interactive constable on patrol system. In such a system, Internet-enabled devices and a mobile application that delivers policing capabilities to front-line officers (to make the work of the force more efficient and appropriate) plays an important role. Such an application improves knowledge exchange, communication practices, and analysis of information within the police force. To achieve this goal, in this paper, we present a scalable and extensible IoT-Enabled Process Data Analytics Pipeline (namely *iProcess*) to enable analysts to ingest data from IoT devices, extract knowledge from this data and link them to process (execution) data. To enhance the real-time dashboard, as a future work, we are working on a novel Platform-as-a-Service that makes it easy for developers of all skill levels to use machine learning technology, the way people use spreadsheet.

References

1. Bandyopadhyay, D., Sen, J.: Internet of Things: applications and challenges in technology and standardization. Wirel. Pers. Commun. **58**(1), 49–69 (2011)
2. Beheshti, A., Benatallah, B., Nezhad, H.: ProcessAtlas: a scalable and extensible platform for business process analytics. Softw. Pract. Exper. **48**(4), 842–866 (2018)
3. Beheshti, A., Benatallah, B., Nouri, R., Chhieng, V.M., Xiong, H., Zhao, X.: Coredb: a data lake service. In: Proceedings of the 2017 ACM on Conference on Information and Knowledge Management, CIKM 2017, Singapore, 06–10 November 2017, pp. 2451–2454 (2017)
4. Beheshti, A., Benatallah, B., Nouri, R., Tabebordbar, A.: CoreKG: a knowledge lake service. In: Proceedings of the VLDB Endowment (PVLDB 2018), vol. 11(12) (2018). https://doi.org/10.14778/3229863.3236230
5. Beheshti, S., Benatallah, B., Motahari-Nezhad, H.R.: Scalable graph-based OLAP analytics over process execution data. Distrib. Parallel Databases **34**(3), 379–423 (2016)
6. Beheshti, S., Benatallah, B., Nezhad, H.R.M.: Enabling the analysis of cross-cutting aspects in ad-hoc processes. In: CAiSE, pp. 51–67 (2013)
7. Beheshti, S.-M.-R., Benatallah, B., Motahari-Nezhad, H.R., Sakr, S.: A query language for analyzing business processes execution. In: Rinderle-Ma, S., Toumani, F., Wolf, K. (eds.) BPM 2011. LNCS, vol. 6896, pp. 281–297. Springer, Heidelberg (2011). https://doi.org/10.1007/978-3-642-23059-2_22
8. Beheshti, S., et al.: Process Analytics - Concepts and Techniques for Querying and Analyzing Process Data. Springer, Cham (2016). https://doi.org/10.1007/978-3-319-25037-3
9. Beheshti, S., Tabebordbar, A., Benatallah, B., Nouri, R.: On automating basic data curation tasks. In: WWW (2017)
10. Benson, D.: The police and information technology. In: Technology in Working Order: Studies of Work, Interaction, and Technology, pp. 81–97 (1993)
11. Bhattacharya, K., Gerede, C.E., Hull, R., Liu, R., Su, J.: Towards formal analysis of artifact-centric business process models. In: BPM, pp. 288–304 (2007)
12. Braga, A.A., Weisburd, D.L.: Police innovation and crime prevention: lessons learned from police research over the past 20 years (2015)

13. Carey, M.J., Onose, N., Petropoulos, M.: Data services. Commun. ACM **55**(6), 86–97 (2012)
14. Casati, F., Castellanos, M., Dayal, U., Salazar, N.: A generic solution for warehousing business process data. In: Proceedings of the 33rd International Conference on Very Large Data Bases, pp. 1128–1137. VLDB Endowment (2007)
15. Da Xu, L., He, W., Li, S.: Internet of Things in industries: a survey. IEEE Trans. Industr. Inf. **10**(4), 2233–2243 (2014)
16. Gerede, C., Su, J.: Specification and verification of artifact behaviors in business process models. In: ICSOC, pp. 181–192 (2007)
17. Kuo, J.: A document-driven agent-based approach for business processes management. Inf. Softw. Technol. **46**(6), 373–382 (2004)
18. Motahari-Nezhad, H.R., Saint-Paul, R., Casati, F., Benatallah, B.: Event correlation for process discovery from web service interaction logs. VLDB J. Int. J. Very Large Data Bases **20**(3), 417–444 (2011)
19. Ngu, A.H.H., Gutierrez, M.A., Metsis, V., Nepal, S., Sheng, Q.Z.: IoT middleware: a survey on issues and enabling technologies. IEEE Internet Things J. **4**(1), 1–20 (2017)
20. Reijers, H., Rigter, J., Aalst, W.: The case handling case. Int. J. Cooperative Inf. Syst. **12**(3), 365–391 (2003)
21. Schonenberg, H., Weber, B., van Dongen, B.F., van der Aalst, W.M.P.: Supporting flexible processes through recommendations based on history. In: BPM, pp. 51–66 (2008)
22. Sun, Y., Song, H., Jara, A.J., Bie, R.: Internet of Things and big data analytics for smart and connected communities. IEEE Access **4**, 766–773 (2016)
23. Sun, Y., Su, J., Yang, J.: Universal artifacts: a new approach to Business Process Management (BPM) systems. ACM Trans. Manage. Inf. Syst. **7**(1), 3:1–3:26 (2016)
24. van der Aalst, W., et al.: Process mining manifesto. In: Daniel, F., Barkaoui, K., Dustdar, S. (eds.) BPM 2011. LNBIP, vol. 99, pp. 169–194. Springer, Heidelberg (2012). https://doi.org/10.1007/978-3-642-28108-2_19

Storytelling Integration of the Internet of Things into Business Processes

Zakaria Maamar[1](✉), Mohamed Sellami[2], Noura Faci[3], Emir Ugljanin[4], and Quan Z. Sheng[5]

[1] Zayed University, Dubai, UAE
zakaria.maamar@zu.ac.ae
[2] ISEP, Paris, France
[3] Université Lyon 1, Lyon, France
[4] State University of Novi Pazar, Novi Pazar, Serbia
[5] Macquarie University, Sydney, Australia

Abstract. This paper discusses the integration of Internet of Things (IoT) into Business Processes (BPs). To define the business logic of thing-aware BPs, existing approaches extend traditional workflow languages (i.e., who does what, why, when, and where) with constructs like things' roles. However, this way of defining the business logic restricts things' operations and, thus, hinders them from initiating ad-hoc/opportunistic collaboration with peers. To overcome this limitation, we tap into the *storytelling* principles to introduce the concept of Process of Things (PoT) as a new way of integrating IoT into BPs. A PoT is specified as a story whose script indicates the characters that things will play as well as the scenes that will feature these things. A PoT, also, allows things to collaborate by offering value-added services to end-users. For demonstration purposes, a hospital scenario is implemented using a combination of real and simulated sensors along with different IoT technologies and communication protocols.

Keywords: Business process management · Internet of Things
Storytelling · Healthcare

1 Introduction

The Internet of Things (IoT), smart cities, wearable devices, and virtual reality/augmented reality are examples of ICT buzzwords that are making the boundaries between reality and fiction vanish. Mark Weiser argues that "...*The most profound technologies are those that disappear. They weave themselves into the fabric of everyday life until they are indistinguishable from it*" [16]. In this paper, we analyze the particular weave (or integration) of IoT ([10], "things" for short) into the know-how of enterprises usually referred to as *Business Process* (BP). "... *A process is nothing more than the coding of a lesson learnt in the past, transformed into a standard by a group of experts and established as a mandatory flow for those who must effectively carry out the work*" [21].

© Springer Nature Switzerland AG 2018
M. Weske et al. (Eds.): BPM Forum 2018, LNBIP 329, pp. 127–142, 2018.
https://doi.org/10.1007/978-3-319-98651-7_8

Existing approaches for integrating IoT into BPs [17,18] usually extend traditional workflow languages with constructs (e.g., things' roles) to define the business logic of thing-aware BPs. However, this way of defining the business logic restricts things' operations and, thus, hinders them from engaging in ad-hoc/opportunistic collaboration with peers. To overcome this limitation, we tap into the *storytelling* [5,25] to introduce the concept of Process of Things (PoT)[1] as a new way of integrating IoT into BPs. Storytelling would allow identifying things according to their capabilities, supporting things to take over new/adjust their capabilities (i.e., mutation), facilitating the (dis)connection of things together through pre-defined relations, and, finally, incentivizing/penalizing things in response to their constructive/destructive participation in processes.

PoT is a new way of capitalizing on IoT opportunities. Our objective is to ensure that things do not function as silos but contribute collectively to offering value-added services to end-users. This could happen by, for instance, identifying relations between things upon which these things will develop *networks* of contacts. A thing uses these networks to reach out to other candidate things for possible inclusion in processes, to avoid conflicts with things prior to their inclusion in processes, and to support its collaboration with things. Atzori et al. develop a social Internet of Things over such relations [3] and stress out the importance of *"exploiting social relationships among things, not among their owners"* [2]. According to Khan et al., these things will constitute a social collaborative IoT environment [12].

To set up a PoT, we analyze things from two perspectives: *capability* (thoroughly discussed in the paper) and *compatibility*. On the one hand, *capability* prescribes a thing's duties once it becomes functional and thus, ready to act collaboratively with other things during the set up of a PoT. Capabilities include sensing, storing, processing, and diffusing with the option of combining them (e.g., sensing and processing). We refer to these capabilities as *individual* and also suggest *group* capabilities (e.g., persuasion and negotiation) that call for the concurrent involvement of multiple things in achieving these capabilities. In this work, we focus on individual capabilities, only. On the other hand, *compatibility* indicates a thing's concern with the participation of other things in the same PoT. This could be established based on things' preferences when working with others so that risks of conflicts are mitigated and/or avoided.

In this paper, we exemplify PoT with a hospital scenario. Various things contribute to this scenario including those related to medical equipment (e.g., thermometers), ambient facilities (e.g., air sensors), and patients (e.g., smart wrists). Making things participate in the same process would require identifying them according to their capabilities, ensuring their collaborative grouping without raising any conflicts, and ensuring their smooth connection. Our contributions are manifold: (*i*) definition of PoT, (*ii*) definition of things' capabilities so, that, they are discovered, (*iii*) definition of relations between things, (*iv*) specification

---

[1] PoT is different from thing-aware BPs as discussed in [11,24]. It is about thing *versus* task as a process's constituent.

and development of PoT using storytelling principles, and (*v*) demonstration of PoT through a case study and an implemented system.

The rest of this paper is organized as follows. Section 2 defines the concept of storytelling, compares process of things to process of tasks, and discusses some related work. Section 3 presents the PoT development approach in terms of concepts and stages. A running example is, also, discussed in this section. The implementation of this example is detailed in Sect. 4. Finally, concluding remarks and future work are presented in Sect. 5.

2 Background

This section, first briefly describes the concept of storytelling and how it is used in certain domains, then compares PoT to process of tasks, and, finally, contrasts the social Internet of Things to the Internet of social things aiming at illustrating examples of social relations among things.

2.1 Storytelling in Brief

Storytelling has been used in different domains such as computer games and educational virtual environments (e.g., [4,5]). Fisher was among the first to propose storytelling to capture life events as a series of ongoing narratives [9]. Storytelling has one main element, *story*, that features the following components [26]: (*i*) *script* that outlines a sequence and/or branching of actions and events related to the *story*, (*ii*) *characters* that set personalities along with their mental attitudes and relationships, and (*iii*) *settings* (*aka scenes*) that include spatio-temporal locations along with objects that *characters* manipulate when they join the *settings*. To develop stories for games, Crawford discusses three ways of adopting storytelling [6]:

- Environment-based: The storywriter establishes a *script* from which the player, who is a *character*, cannot deviate. This ensures that the player's actions lead to developing a coherent sequence of *scenes*. The environment-based storytelling is appropriate when stories require strict compliance with the *script*.
- Data-driven: The storywriter provides the player a set of generic *story* components so, that, she combines them together. This allows the player to develop personal stories. The data-driven storytelling is appropriate when changes during script execution are known in advance, so, that, initial scripts are adapted.
- Shared-authoring: The player shapes and constrains the story's scope by working out the story's other possibilities. This fosters the player's autonomy and further collaboration among players. The shared-authoring storytelling is appropriate when changes during script execution are not known in advance, so, that, initial scripts can be expanded.

2.2   Process of Things *versus* Process of Tasks

The following highlights a process of things' similarities and differences to a regular process of tasks. There are some similarities:

- Just as a process of tasks has a business logic defining who does what, when, and where, a PoT will have a *story* that identifies in an abstract way the necessary things along with their capabilities and possible connections to other things. More on capabilities and connections are, also, given in Sect. 3.3.
- Like dependencies between tasks in processes, a PoT will have connections build upon capability-driven relations between things such as complementary, antagonism, and competition. More details on relations are given in Sect. 3.3.
- Both process of tasks and PoT require computing, storage, and/or communication resources at run-time.

The two, also, have some differences:

- Contrary to a process of tasks whose runtime instantiation leads to several process instances with the same set of tasks, instantiating the same PoT several times could call for different things depending on their capabilities and current availabilities. Thing participating in a certain PoT instance might be different in another instance of the same PoT (although the same story is used).
- Unlike a process of tasks where the tasks are known in advance, a PoT will have a set of *core* things (also known in advance) and a set of *optional* things that are part of these core things' networks of contacts (using capability-driven relations). Upon the recommendations of core things, optional things are added to a PoT subject to end-user's approval of the additional cost (if any). Needless to say, dropping a core thing from a PoT results in dropping all of its optional things.
- Unlike a process of tasks, a PoT is autonomic. In fact, executing a task requires a third party (that could be a human). Thing is self-managed needing a trigger, only, to start functioning.

To wrap-up, we consider a PoT as a process in the sense that there is a business logic and a set of tasks. This business logic is mapped onto a script and the set of tasks onto actions that characters (i.e., things) will implement. In this paper, we discard the use of BPM terminology like modeling and enactment for the sake of complying with the terminology of storytelling.

2.3   Related Work

Like many ICT fields, IoT has ridden the Web 2.0 wave resulting into terms like social Internet of Things [3,27] and Internet of Social Things [20]. Other terms, associated with this wave, include social cloud [8], social business process [13], and social Web service [14], etc. As we align PoT with the concept of social things, this section discusses some works on blending social computing with IoT.

In [3,19], Atzori et al. treat things as intelligent objects and suggest that models designed for studying social networks of humans can be extended to social networks of objects. These networks could be built upon relations like parental (similar objects built in the same period by the same manufacturer), co-location (objects in the same venue), co-work (objects participating in the same scenario), ownership (objects having the same user), and social (when objects come into contact sporadically or continuously). Atzori et al. note the paradigm shift from human-object interaction to object-object interaction. In [15], Maamar et al. highlight the tremendous opportunities that both wireless technologies and mobile devices offer so, that, ad-hoc collaboration between independent messengers (unexpectedly) takes shape. A messenger is a software agent running on a user's mobile device and is in charge of conveying data (e.g., Web services' descriptions) from one repository to another. When mobile devices are in the "vicinity" of each other, they form mobile ad hoc collaborative environment without any pre-existing communication infrastructure.

In [17], Meroni discusses the integration of IoT into business process management and notes that this integration faces problems of process compliance and smart object configuration. From an IoT perspective, smart objects are devices that support decentralize data computation and acquisition. To this end, a smart object is equipped with a sensor network, a single board computing unit, and a communication interface. Meroni enriches business process modeling notations with constructs that depict smart objects' roles and needs inside BPs. In line with the work of Meroni, Meyer et al. examine the integration of IoT devices into BPs as resources [18]. IoT devices like sensors interact with the physical environment and hence, could feed processes with relevant, live data. However, Meyer et al. note that taking into account IoT devices' characteristics during process modeling is not properly handled. To this end, they suggest mapping the main abstractions and concepts of the IoT domain (namely IoT service, physical entity, IoT device, and native service) onto specific notations and constructs.

The aforementioned efforts illustrate the ICT community's interest in blending social computing with IoT. Despite this interest, a good number of challenges and open issues that result from this blend remain untackled according to Ortiz et al. [22]. This includes defining a social thing architecture, addressing interoperability of things, discovering things, managing energy consumption of things, handling security, privacy, and trust of things, etc. Our work tackles another challenge by examining the on-the-fly combination of things. A PoT will have an operation model that inventories things according to a particular context (e.g., meeting room) and capitalizes on relations between things to add more unplanned things to this model.

3   Process-of-Things Development Approach

In this section, we provide an overview of the approach for developing PoTs and then, discuss this approach's *preparatory*, *design*, and *shooting* stages.

3.1   Definitions

To formally define PoT, we deem necessary to identify its components using the *storytelling* basics of Sect. 2.1. A PoT is a story defined by, at design-time, an abstract *script* and, at run-time (which we refer to as "shooting" time), a set of concrete *scenes* (or *settings*) that are the result of instantiating the script.

On the one hand, a script has one element that is (main) goal. Depending on the goal complexity, it is recursively decomposed until terminal goals are obtained[2]. Each terminal goal is defined with the following elements: (*i*) name, (*ii*) who will achieve it in terms of roles-to-play and/or capacities-to-have along with the necessary actions to take over things, (*iii*) when will it be achieved in terms of dates, (*iv*) how will it fit into the chronology of achieving other terminal goals, and (*v*) where will it be achieved in terms of locations. To specify the grammar of a PoT's script, we use Augmented Backus-Naur Form (ABNF) [7] as per Listing 1.

On the other hand, a scene has the following elements: (*i*) purpose (1,1)[3] acting as an identifier, (*ii*) selected characters (0,n) as a subset of all things (1,n) automatically/manually detected[4] in the scene, (*iii*) temporal start, (*iv*) pre-scenes (0,n), (*v*) post-scenes (0,n), and (*vi*) physical space. A character is either a living thing or a non-living thing.

Listing 1. PoT's *script* grammar

```
1   Script      = Goal
2   Goal        = (G:Name *[Goal]) / TerminalGoal
3   TerminalGoal = '('TG:Name ';' *(C:Character[',']) ';' *(T:Time[',']) ';' Chron:
        Chronology';'              *(L:Location[',']) ')'
4   Character   = *((r:Role/c:Capacity)'.'Action[','][','.'objectAction][','])
5   Time        = ((AFTER / BEFORE) Date) / (Time [AND / OR] Time)
6   Chronology  = *([PRE TG:Name][',']) *([POST TG:Name][','])
7   Location    = Place
8   Name        = *(ALPHA / DIGIT)
9   objectAction = ('C:r:'Role/'C:c:'Capacity)
10  Role        = *(ALPHA / DIGIT)
11  Capacity    = *(ALPHA / DIGIT)
12  Date        = *(ALPHA / DIGIT)
13  Time        = *(ALPHA / DIGIT) / currentTime
14  Action      = *(ALPHA / DIGIT)
15  Place       = *(ALPHA / DIGIT) / currentLocation
```

At the shooting time, binding the elements of a script's terminal goal to the elements of a scene occurs as follows: name (`TG:Name`) corresponds to purpose, roles-to-play/capacities-to-have ((`r:Role/c:Capacity).Action.objectAction`) identify selected characters along with their actions, date (`T:Time`) corresponds to temporal start, chronology of terminal goals (`Chron:Chronology`) defines pre- and post-scenes, and, finally, location (`L:Location`) corresponds to physical space. Listing 2 is the ABNF grammar of a PoT's scene:

---

[2] Readers are referred to [23] for more details on goal decomposition.

[3] (*x*, *y*) refers to min and max occurrence.

[4] Detection refers to thing availability in a *scene* from which certain things are selected for inclusion in a *script* completion based on roles/capacities' requirements. Not all detected things would be relevant for a *scene*.

Listing 2. PoT's *scene* grammar

```
1   Scene    = '('purpose':'sceneID';'
2                temporalStart':'*(ALPHA / DIGIT)';'
3                physicalSpace':'*(ALPHA / DIGIT)';'
4                selectedCharacters':'*(characterID):'('?'[r:Role]'?'[',']')/('?'[c:Capacity]
    '?'[',']')';'
5                //?x? means missing character from the scene
6                preScenes':'*([sceneID][',']')';' //sceneID differs from the purpose's
    sceneID
7                postScenes':'*([sceneID][',']')';'//sceneID differs from the purpose's
    sceneID
8   sceneID   = *(ALPHA / DIGIT)
9   characterID = *(ALPHA / DIGIT)
```

In Listing 2-line 6, the comment ($//?x? \cdots$) is about the absence of a character that is necessary for completing a scene instantiation. This absence along with PoT's script and scenes illustration are exemplified in the next section.

3.2   An Example

To exemplify the script and scenes of a PoT, we use the example of a hospital that is on high-alert being close to a major car accident. The hospital has different state-of-the-art equipment and facilities that showcase how IoT can smoothen operations and improve efficiency. Wards have ambient sensors for temperature automatic-control, life-support machines have RFID tags for better tracking and maintenance, and smart wrists allow real-time transmission of patients' vitals to appropriate recipients.

Let us consider an injured driver who requires an immediate surgery due to brain bleeding. Unfortunately, all the hospital's operating theaters are occupied forcing the medical team to improvise one of the emergency ward's rooms as an operating theater[5]. As a result, the relevant PoT is activated as per the hospital's prescribed guidelines that would ensure surgery success and medical staff's and equipment's availability. To begin with, the PoT's script refers to *stop-brain-bleeding* goal whose design-time decomposition leads to several other goals such as *prepare-patient, perform-surgery, diagnose-patient, give-medication,* and *prepare-operating-theater* (Fig. 1). Some of these goals are terminals and, thus, will be instantiated through concrete scenes at shooting-time.

In compliance with Listing 1, we detail the script for *stop-brain-bleeding* in Listing 3. This script's main goal `G:stop-brain-bleeding` (line 1) is decomposed into `G:prepare-patient` and `G:perform-surgery`. Each is also decomposed into two terminal goals (i.e., `TG:diagnose-patient` (lines 3–7) and `TG:give-medication` (line 8) connected to `G:prepare-patient`). The terminal goal `TG:diagnose-patient` has a name (line 3), characters defined by their roles/capabilities and actions to execute (e.g., nurse takes temperature of a patient (`C:r:nurse.takeTemp.C:r:patient`) or (`OR`) take temperature using a connected thermometer (`C:c:thermometer.takeTemp.C:r:patient` (line 4)), time and location, when and where the script should be executed

---

[5] Improvisation might raise concerns with the availability of some necessary things as prescribed in a scene (Listing 2-line 7).

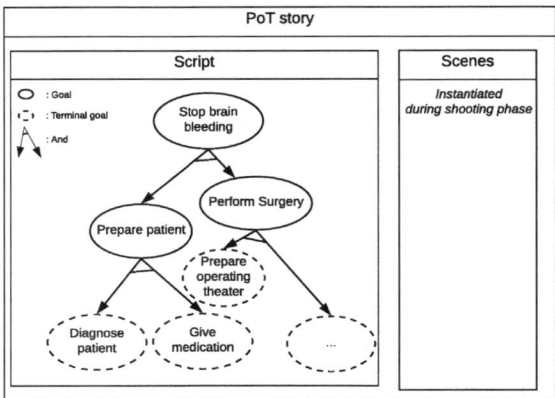

Fig. 1. Partial representation of the script for stop-brain-bleeding surgery

(T:currentTime/L:currentLocation (line 7)), and chronology (i.e., after the terminal goal TG:give-medication (line 7)).

Listing 3. Example of *script* definition

```
1    Script   = G:stop-brain-bleeding
2              G:prepare-patient
3              (TG:diagnose-patient;
4               C:r:nurse.takeTemp.C:r:patient OR C:c:thermometer.takeTemp.C:r:patient,
5               C:r:nurse.takePress.C:r:patient OR C:c:smartWrist.takePress.C:r:patient,
6               C:r:doctor.diagnose.C:r:patient;
7               T:currentTime; Chron: POST TG:give-medication; L:operatingTheater)
8              (TG:give-medication; ..... )
9              G:perform-surgery
10             (TG:prepare-operating-theater; .....)
11             (TG:.....)
12             ...
```

In compliance with Listing 2, we detail a scene associated with the terminal goal TG:diagnose-patient in Listing 4. This scene's elements are as follows: (*i*) temporalStart is the time of the scene instantiation (currentTime), (*ii*) physicalSpace is the emergency ward room (currentLocation //emergencyRoom1) since the hospital's operating theaters are occupied, (*iii*) selectedCharacters are the nurse (Miranda), the patient's smart wrist (sw1), and the doctor who is not available (?r:doctor?) (in this case a doctor needs to be identified), and (*iv*) preScenes and postScenes refer to null and give-medication, respectively.

Listing 4. Example of *scene* instantiation

```
1    Scene = (purpose: diagnose-patient;
2            temporalStart: currentTime // Thu May 18 2017 17:24:06 GMT+0200;
3            physicalSpace: currentLocation //emergencyRoom1;
4            selectedCharacters: Miranda:r:nurse, sw1:c:smartwrist, ?C:r:doctor?;
5              preScenes: null;
6              postScenes: give-medication)
```

Out of Listing 4, it is worth making two comments that will be handled in Sect. 3.3:

- In Line 3, `currentLocation` refers to `emergencyRoom1` that is different from the operating theater stated in Line 7 of Listing 3 that should be an operating room. This difference in location raises concerns with the availability of necessary things.
- In Line 4, `?C:r:doctor?` highlights the case of a missing character from a scene as stated in line 6 of Listing 3.

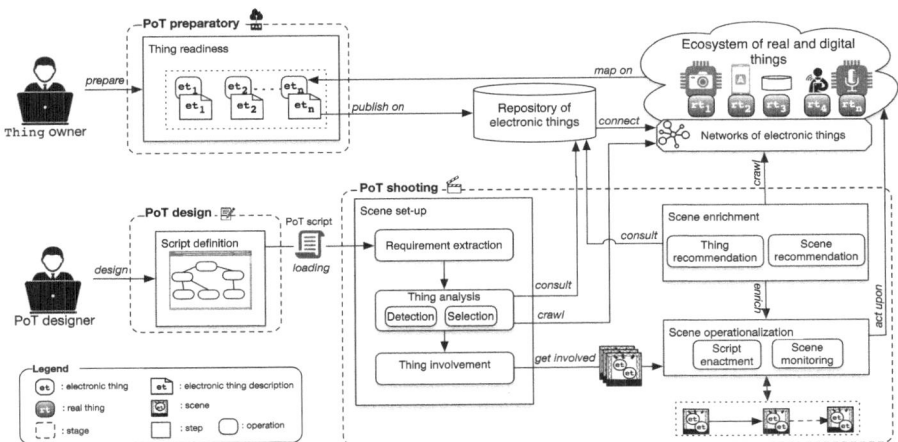

Fig. 2. Stages associated with PoT development

3.3   Development Stages

PoT development revolves around 3 stages: *preparatory* that handles (real and digital) thing readiness in terms of mapping, description, and connection, *design* that handles script definition, and finally *shooting* that handles scene management in terms of set-up, operationalization (put into action), and enrichment. The chronology of performing these stages is illustrated with Fig. 2 along with the necessary stakeholders, namely, ecosystem of real and digital things, things' owners, and PoT designers. Some of the salient features of the IoT development stages include *customization* (i.e., things' changing and various capabilities are taken into account when meeting a PoT's needs), *openness* (i.e., things can join and leave the ecosystem without impacting PoT definition (happens at the abstract level)), *flexibility* (i.e., PoT operationalization can be adjusted thanks to enrichment), and *context-awareness* (i.e., things that join PoT are detected when needed and not designated upfront). The rest of this section will present the details of the respective 3 stages.

Preparatory Stage. The owners of things take part in the preparatory stage by mapping real (e.g., thermometer and patient) and digital (e.g., patient medical record) things onto electronic things, describing these electronic things with focus on their capabilities, publishing these descriptions on a dedicated repository, and last but not least developing networks of electronic things upon certain capability-driven (social) relations.

1. Thing mapping ensures that real and digital things have an electronic presence in the virtual world. According to Atzori et al., *"in the IoT, everything real becomes virtual, which means that each person and thing has a locatable, addressable, and readable counterpart on the Internet. These virtual entities can produce and consume services and collaborate towards a common goal"* [1]. Electronic things act as proxies during the detection, selection, and involvement of real and digital things in PoTs. These 3 actions happen during the shooting phase and depend on scenes' requirements like necessary things (or characters as per *script* terminology) for executing actions (e.g., taking patient's temperature).

2. Thing description supports matching things' capabilities to PoTs' requirements so, that, necessary things are selected (Listing 2, Line 4). For the sake of compliance with existing standards and practices, we adopt the WoT (Web of Things) Current Practices (CP)[6] recommended by W3C to describe things in terms of *semantic metadata, security, communication, and interaction resources*. A thing description can be embedded into the thing or hosted somewhere on the Web. This ensures that available descriptions can be applied to existing things and can complement IoT platforms with rich metadata enabling across platform interoperability.

3. Network of thing development supports the enrichment of PoTs' scripts by developing capability-driven (social) relations between things[7]. These relations are in line with Atzori et al.'s 5 relations (though the social dimension in Atzori et al.'s work is not stressed-out) defined as follows [19]: *parental, co-location, co-work, ownership,* and *social.* Our social relations go beyond the notion of contact. They allow things to be aware of peers from 3 perspectives: recommendation exemplified with complementarity relation, opposition exemplified with antagonism relation, and exclusion exemplified with competition relation (below *et* stands for electronic thing).

   (a) Complementary(et_i, et_j) defines the concurrent participation of things in joint PoTs, e.g., et_i:temperatureSensor and et_j:humiditySensor; et_i could recommend to a PoT designer that et_j participates in this PoT.

   (b) Antagonism(et_i, et_j) defines the "sensitivity" (or "friction") that exists among things when both participate in joint PoTs, e.g., et_i:mp3Player and et_j:dvdPlayer; both deliver a certain form of entertainment.

   (c) Competition(et_i, et_j) defines the exclusion among things as only one can participate in a PoT, e.g., either t_i:spareBed or t_j:inUseBed.

---

[6] w3c.github.io/wot/current-practices/wot-practices.html.

[7] Connecting things allows to address the lack of/missing characters from a scene (Listing 2, Line 4).

We use the aforementioned relations to develop networks of electronic things (Fig. 2). These networks' edges provide useful insights during the development of PoTs. Due to limited space, only the antagonist network is presented. Each time et_i and et_j participate in a joint PoT in case of missing characters, the edge's weight in the network is reassessed if the edge already exists. Otherwise, a new edge connecting et_i and et_j together is set-up (Eq. 1). A high value reflects a strong co-presence between things.

$$w_{antagonism(et_i,et_j)} = \frac{jointPoT(et_i,et_j)}{participatedPoT(et_i|\neg et_j) + participatedPoT(\neg et_i|et_j)} \tag{1}$$

where: $jointPoT(et_i, et_j)$ is the number of times et_i and et_j participated together in joint PoTs and $participatedPoT(et_i \mid \neg et_j)$ is the total number of times et_i took part in PoTs without et_j and *vice versa*.

Design Stage. At this stage, PoT designers work out the necessary elements of a script (Sect. 3.1). The objective is to ensure that the decomposition of each goal is complete in terms of who will do what, when, and where.

Our assumption is that PoT designers may not be familiar with how to define a script using ABNF grammar. Thus, we have developed a tool that would make this grammar transparent to PoT designers. The tool generates a script's textual description that is compliant with the proposed grammar (Listing 1) from a script's graphical representation (i.e., goal decomposition tree). A PoT designer builds a root node denoted as the script's main goal and fulfills specific node attributes such as the number of children referring to subgoals along with their type (i.e., goal or terminal goal). In turn, the tool connects the root node to its child nodes. For each node, the PoT designer describes its elements as per its type. For instance, only a terminal-goal has characters and chronology as attributes. Upon tree building completion, the tool takes the goal decomposition tree including the node attributes as input and relies on the grammar to generate the script's textual representation.

Shooting Stage. It is a cornerstone to PoT development by performing 3 scene-related operations that are *set-up*, *operationalization*, and *enrichment* (the last two could happen concurrently). Upon loading a PoT's script during the set-up operation, the terminal goals that result from decomposing the script's (main) goal, identify the necessary scenes that will drive the PoT shooting.

Scene set-up involves three modules (Fig. 2): *requirement extraction*, *thing analysis*, and *thing involvement*.

– The *requirement-extraction* module identifies the terminal goals' requirements in terms of character types (e.g., nurse for roles/scanner for capacities), time (i.e., time frame/interval (e.g., in the morning) or point in time (e.g., current time)), location (e.g., operating theater), and chronology of scenes (e.g., check patient's vitals then assign a priority to the case). All these requirements are included in a script definition (Listing 1).

- The *thing-analysis* module detects manually and/or automatically the things that are currently present in the under-preparation scene[8] so, that, they are selected after matching their capabilities to the requirements of the scene's roles/capacities. The selection could take advantage of the different networks for instance, antagonist to avoid conflicts between future characters to include in a scene and competition to identify the best future characters in a scene. Capability/requirement matching could be unsuccessful indicating that some (or may be all) present things (i.e., detected) are not relevant for a scene and/or necessary things (to become characters) for a scene are not available (e.g., Listing 4, line 4, `?C:r:doctor?`). In the first case, the things are discarded from the scene. In the second case, the repository of things is consulted to identify the things that could either take over the roles in the scene or fulfill the capacities in the scene. Assuming that things are successfully identified, the next action consists of checking the competition network to select the best things and then the antagonist network to ensure that the new characters associated with the best selected things do not conflict with any existing character that is already included in the PoT and/or scene. Upon confirmation of the new characters, the scene's characters could recommend more things as per the description of the scene enrichment operation detailed below.
- Finally, the *thing-involvement* module confirms the participation of all the necessary characters in the scene prior to launching the scene operationalization operation.

Scene operationalization involves two modules (Fig. 2): *script enactment* and *scene monitoring*.

- The *script-enactment* module executes the script by acting upon real and digital things. During the scene operationalization operation, the elements of a scene's as per Listing 2 are instantiated for instance, a date and time are assigned to the scene's current time (e.g., Listing 4, line 2, `Thu May 18 2017 17:24:06 GMT+0200`) and the real-world setting where the shooting is happening, is assigned to the location's name of the *scene* (e.g., Listing 4, line 3, `emergencyRoom1`).
- The *scene-monitoring* module tracks and analyzes the events that could undermine a script smooth execution like characters leaving *a scene* suddenly (e.g., breakdown) or intentionally (e.g., withdrawal by their owners). This module, also, reports these events to the *scene set-up* operation that is activated again to ensure the continuity of the script execution by replacing characters using the competition network while ensuring the compatibility of all characters using the antagonism network.

Scene enrichment involves two modules (Fig. 2): *thing recommendation* and *scene recommendation*.

- The *thing-recommendation* module targets a specific scene by considering the possible inclusion of more things in a scene. To this end, the support

---

[8] Detection does not fall into the scope of this work.

of network of things is sought as per the complementary, competition, and antagonist relations. For instance, the complementary network provides the PoT designer with things recommended by characters in the scene either upon his request or when the edge's weight exceeds some threshold. This enables to add new characters to the PoT after the designer's approval. As many things can be recommended, the competition network can select the best ones. Finally, the antagonist network enables to check if all characters (existing and recommended) are not in conflict before updating the scene.

– The *scene-recommendation* module deals with adding more scenes to a script and/or dropping some from a script. Both cases are part of future work.

4   Implementation

To demonstrate the technical feasibility of our PoT development approach, we used multiple technologies and protocols to implement a system for the hospital scenario following a layered architecture (Fig. 3). This system is deployed on a Linux Apache server and summarized with a short video available at https://social.connect.rs/pot/video.mp4. From a development perspective, we used Python, PHP, and MySQL for the back-end and HTML, CSS, JavaScript, Bootstrap library, and Paho JavaScript Client library for the front-end. From a communication perspective, we used Hyper Text Transfer Protocol (HTTP) and Message Queue Telemetry Transport (MQTT) protocol.

As per Sect. 3.3, the *preparatory stage* consists of making things (in fact characters such as Michael as General Practitioner, Angelina as Nurse, and Thermometer-1 as Thermometer) ready for inclusion in future PoTs. To this end, the hospital's PoT designer describes the necessary things using JSON that is, afterwards, stored into the repository of electronic things. Thing descriptions are created through dedicated GUIs. The repository of electronic things is implemented as a relational database containing various real-time details on things such as accessibility, capability, and communication protocol. For implementation needs, we used three real non-living (two temperature and humidity sensors (one DHT11 with NodeMCU v3 module and another DHT21 with Wemos D1 module) and one Arduino Uno with esp8266 and RFID-RC522), and three other simulated non-living things deployed by using Bevywise IoT simulator (www.bevywise.com/iot-simulator.). MQTT protocol supports the communication between the system and all things with the open source Mosquito broker being part of the hospital system. This broker allows, first, living things to notify the system about their availabilities using RFID tags, and second, non-living things to publish various messages. Messages are sent to an in-house Python module, acting as a gateway, along with MQTT topics such as location and device ID. Message content formatted in JSON also carries other details such as sensor readings and timestamps.

As per Sect. 3.3, during the *design stage*, the PoT designer defines scripts using a dedicated GUI as per the grammar of Listing 1. For the PoT's terminal goals, the designer defines the necessary roles (for living things) and capacities

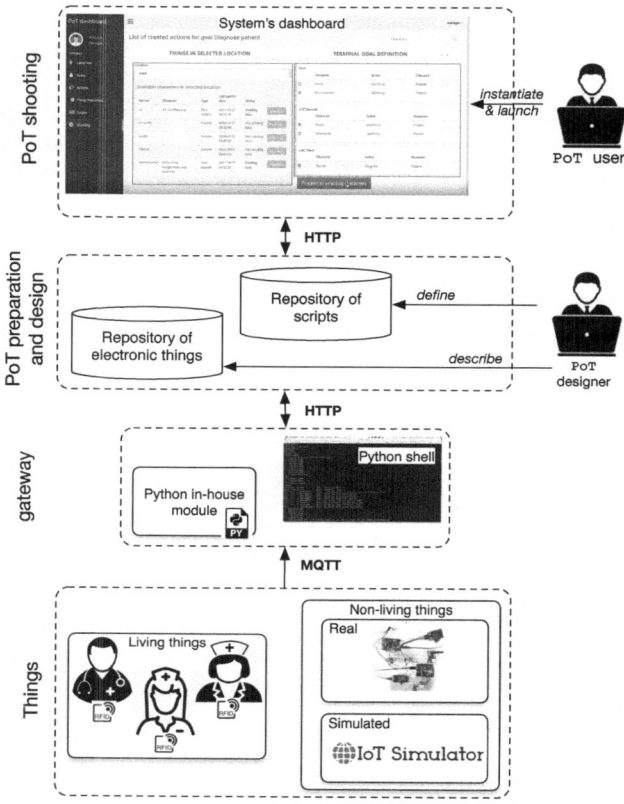

Fig. 3. Layer-representation of the system architecture

(for non-living real an simulated things) referring to some actions to be executed over certain objects. Examples of actions and objects are mentioned in Listing 2.

As per Sect. 3.3, during the *shooting stage*, the PoT user such as nurse Angelina selects one of the pre-defined scripts depending on the current situation like stop-brain-bleeding surgery (Fig. 1). Upon identification of the script's terminal goals, the scene associated with each terminal goal is instantiated by for instance, selecting the necessary characters and stating the current location and time. The system automatically parses the repository of electronic things to seek for all available things that act as characters for inclusion in that scene.

5   Conclusion

We have presented an approach for designing and developing *process of things* (PoT). By analogy to process of tasks that has a business logic and executors, a PoT is defined with a story having a script (what to say and what to do), a set of characters (who will act in the script), and a set of scenes (where to shoot the script). At run-time, real things (e.g., sensors and doctors) fill in for characters and places (e.g., meeting rooms) fill in for scenes. Three stages (preparatory, design, and shooting) have been recommended for developing PoTs, contributing each to defining scripts, characters, and scenes. To demonstrate PoT technical feasibility, a hospital scenario has been implemented using different technologies and communication protocols of the Internet of Things (IoT). In the future, we will experiment the approach on a real case-study. Also, we would like to define the principles and foundations of the *social Internet of "social things"*. On the one hand, things in the social Internet are configured and controlled in preparation for their integration into networks developed upon certain relations like those of Atzori et al. [3,19]. The social Internet will provide the necessary protocols for setting-up, managing, and maintaining the networks of things. On the other hand, a social thing will be empowered with the necessary capabilities for "crawling" the networks of things looking, for instance, partners, avoiding partners, and forming alliances with partners.

References

1. Atzori, L., Carboni, D., Iera, A.: Smart things in the social loop: paradigms, technologis, and potentials. Ad Hoc Netw. **18**, 121–132 (2013)
2. Atzori, L., Iera, A., Morabito, G.: SIoT: giving a social structure to the Internet of Things. IEEE Commun. Lett. **15**(11), 1193–1195 (2011)
3. Atzori, L., Iera, A., Morabito, G., Nitti, M.: The Social Internet of Things (SIoT) - when social networks meet the Internet of Things: concept, architecture and network characterization. Comput. Netw. **56**(16), 3594–3608 (2012)
4. Cavazza, M., Charles, F., Mead, S.J.: Character-based interactive storytelling. IEEE Intell. Syst. **17**(4), 17–24 (2002)
5. Charles, F., Cavazza, M., Smith, C., Georg, G., Porteous, J.: Instantiating interactive narratives from patient education documents. In: Peek, N., Marín Morales, R., Peleg, M. (eds.) AIME 2013. LNCS (LNAI), vol. 7885, pp. 273–283. Springer, Heidelberg (2013). https://doi.org/10.1007/978-3-642-38326-7_39
6. Crawford, C.: Chris Crawford on Interactive Storytelling (New Riders Games). New Riders Games, San Francisco (2004)
7. Crocker, D., Overell, P.: Augmented BNF for Syntax Specifications: ABNF. STD 68, RFC Editor, January 2008
8. Elnaffar, S., Maamar, Z., Sheng, Q.Z.: When clouds start socializing: the sky model. Int. J. E-Bus. Res. **9**(2), 1–7 (2013)
9. Fisher, W.R.: The narrative paradigm: in the beginning. J. Commun. **35**(4), 74–89 (1985)
10. Georgakopoulos, D., Jayaraman, P.P.: Internet of Things: from internet scale sensing to smart services. Computing **98**(10), 1041–1058 (2016)
11. Haller, S., Magerkurth, C.: The real-time enterprise: IoT-enabled business processes. Technical report (2011). www.iab.org/wp-content/IAB-uploads/2011/03/Haller.pdf. Accessed Mar 2018
12. Khan, W.Z., Aalsalem, M.Y., Kha, M.K., Arshad, Q.-A.: When social objects collaborate: concepts, processing elements, attacks and challenges. Comput. Electr. Eng. **58**, 397–411 (2017)

13. Maamar, Z., Faci, N., Kajan, E., Sakr, S., Boukhebouze, M., Barnawi, A.: How to make business processes "Socialize"? EAI Endorsed Trans. Ind. Netw. Intell. Syst. **2**(5), e2 (2015)

14. Maamar, Z., Hacid, H., Huhns, M.N.: Why Web services need social networks. IEEE Internet Comput. **15**(2), 90–94 (2011)

15. Maamar, Z., Mahmoud, Q.H., Derhab, A.: Enabling ad-hoc collaboration between mobile users in the MESSENGER project. Cluster Comput. **10**(1), 67–79 (2007)

16. Mark, W.: The computer for the 21st century. ACM SIGMOBILE Mob. Comput. Commun. Rev. **3**(3), 3–11 (1999)

17. Meroni, G.: Integrating the Internet of Things with business process management: a process-aware framework for smart objects. In: CAiSE 2015 Doctoral Consortium at the 27th International Conference on Advanced Information Systems Engineering (CAiSE 2015), Stockholm, Sweden, pp. 56–64. Springer (2015)

18. Meyer, S., Ruppen, A., Magerkurth, C.: Internet of Things-aware process modeling: integrating IoT devices as business process resources. In: Salinesi, C., Norrie, M.C., Pastor, Ó. (eds.) CAiSE 2013. LNCS, vol. 7908, pp. 84–98. Springer, Heidelberg (2013). https://doi.org/10.1007/978-3-642-38709-8_6

19. Militano, L., Nitti, M., Atzori, L., Iera, A.: Enhancing the navigability in a social network of smart objects: a shapley-value based approach. Comput. Netw. **103**, 1–14 (2016)

20. Nansen, B., van Ryn, L., Vetere, F., Robertson, T., Brereton, M., Dourish, P.: An internet of social things. In: Australian Computer-Human Interaction Conference on Designing Futures: The Future of Design (OzCHI 2014), Sydney, Australia, pp. 87–96. ACM (2014)

21. OpenKnowledge. Social Business Process Reengineering. Technical report (2012). http://socialbusinessmanifesto.com/social-business-process-reengineering. Accessed Mar 2018

22. Ortiz, A.M., Hussein, D., Park, S., Han, S.N., Crespi, N.: The cluster between Internet of Things and social networks: review and research challenges. IEEE Internet Things J. **1**(3), 206–215 (2014)

23. Porteous, J., Cavazza, M., Charles, F.: Applying planning to interactive storytelling: narrative control using state constraints. ACM Trans. Intell. Syst. Technol. **1**(2), 10:1–10:21 (2010)

24. Suri, K., Gaaloul, W., Cuccuru, A., Gerard, A.: Semantic framework for Internet of Things-aware business process development. In: International Conference on Enabling Technologies: Infrastructure for Collaborative Enterprises (WETICE 2017), Poznan, Poland, pp. 214–219. IEEE (2017)

25. Ware, S.G., Young, R.M., Harrison, B., Roberts, D.L.: A computational model of narrative conflict at the fabula level. IEEE Trans. Comput. Intell. Artif. Intell. Games **6**(3) (2014)

26. Young, R.M., Ware, S.G., Cassell, B.A., Robertson, J.: Plans and planning in narrative generation: a review of plan-based approaches to the generation of story, discourse and interactivity in narratives. Sprache und Datenverarbeitung Spec. Issue Form. Comput. Models Narrat. **37**(1–2), 41–64 (2013)

27. Zhang, C., Cheng, C., Ji, Y.: Architecture design for social web of things. In: International Workshop on Context Discovery and Data Mining (ContextDD 2012), Beijing, China, pp. 3:1–3:7. ACM (2012)

Leveraging Unstructured Data to Analyze Implicit Process Context

Renuka Sindhgatta[1(✉)], Aditya Ghose[2], and Hoa Khanh Dam[2]

[1] IBM Research, Bangalore, India
renuka.sr@in.ibm.com
[2] University of Wollongong, Wollongong, Australia
{aditya.ghose,hoa}@uow.edu.au

Abstract. Adapting a business process to different context requires identifying various situations and evolving the process to support such situations. Previous work focused on modeling, observing and collecting contextual information. Furthermore, impact of context on process or resource performance has been studied. However, much of the work considers explicit contextual information that is defined by domain experts. There are several implicit contextual dimensions, that are difficult to model as all situations cannot be anticipated a priori. Context mining involves analysis of process logs to identify context and correlate with process performance indicators or outcomes. In this work, we leverage unstructured data available in user comments or mails to discover implicit context of the process. We automatically analyze textual data and group process instances by applying information extraction and text clustering techniques. Groups of process instances are correlated to their process outcomes to filter irrelevant information. We apply the approach on real-world process logs to identify contextual information.

Keywords: Process context · Natural language processing
Cluster analysis · Process execution logs

1 Introduction

Analyzing and (machine) learning impact of the business process context (or the environmental factors), on its execution helps adapting and improving the process [9]. There exists many interpretations of the notion of context in various disciplines including mobile applications and eCommerce personalization. In one of the early works by Dourish [5], two views of context are presented. First, a *representational view*, where context is defined as information that is stable, can be defined for an activity and is separable from the activity. Hence, context is described using a set of attributes or dimensions. An example of a representational process context is the *hour of the day* when the process executes. It is independent of the activity and yet has an impact on the execution of activity (peak workload). Second, an *interactional view*, where context is dependent on

© Springer Nature Switzerland AG 2018
M. Weske et al. (Eds.): BPM Forum 2018, LNBIP 329, pp. 143–158, 2018.
https://doi.org/10.1007/978-3-319-98651-7_9

the activity, and can be dynamically produced by the activity. An example of interactional view is the non-availability of a customer to confirm the purchase of an insurance claim. While the situation is not a part of the process, it is dynamically created as the activity requires confirmation by the customer.

Business process context modeling considers the representational view, which we term as explicit context: information that is identified by domain experts and can be defined a priori. Saidani et al. [23] define a meta-model of context for a business process. The meta-model comprises of context entity, context attributes and context relationships. A domain expert can define a context model based on the meta-model and the contextual information can be observed from the process execution logs. For example, in the insurance claim process, a domain expert would indicate that the location of customer as contextual information, as the process path and outcome could vary for customers in different locations. These attributes are characterized as *explicit contextual dimensions*. Existing approaches extract contextual dimensions from structured information in process logs, and use supervised learning methods to predict process or resource performance [10, 11, 25, 26].

There are situations that arise as a part of performing a task or an activity (interactional view), and may not be known a priori. These *implicit contextual dimensions* need to be discovered from various sources of information. For example, in an IT application maintenance process, when performing the task of resolving IT problem, the worker or resource may find that, certain legacy applications require more time to resolve as multiple interlinked applications need to be restarted, while a new application using web services takes less time as it requires restart of just that specific web service. This information is implicit and once identified, the process redesign could assign different resolution times based on the new contextual dimension of type of application - legacy application or service based application. The source of identifying the underlying implicit context can be unstructured information available as textual comments that are recorded during the process execution.

In this work, we study the problem of exploiting unstructured textual data to discover implicit context. In the proposed framework, textual data is extracted from execution logs of process instances. Commonly occurring situations are identified by applying text clustering methods. A few relevant clusters are semi-automatically selected by applying filtering rules and choosing clusters with significantly different process performance. The clusters of textual information, can be considered as input to identifying contextual information. This approach helps domain experts discover possible contextual dimensions. To the best of our knowledge, discovery of process context from unstructured or textual data available with process execution histories has not been considered so far. To summarize, the following are the main contributions of our work:

– Introduce the research problem of mining context from textual information available during the process execution.
– Propose a semi-automated approach of identifying context using textual information available in process execution logs.

The paper is organized as follows. Section 2 presents a real-life motivating example, followed by a background of concepts used in our work (Sect. 3). The overall approach is outlined in Sect. 4, and a detailed empirical evaluation is presented in Sect. 5. Related work is presented in Sect. 6, followed by conclusions and future work in Sect. 7.

2   Motivating Example

Table 1, contains textual information logged by workers or resources involved in the process of maintaining IT applications. A problem is reported by a customer. The resource or worker allocated to the task, evaluates the problem, identifies and executes relevant resolution, confirms with the customer if the problem has been resolved. At every step in the process of analyzing and resolving the problem, the details are recorded in an incident management system (process aware information system). Examples in Table 1 are representative of typical challenges with textual logs of business processes: (i) varying informativeness from being very brief to very detailed, (ii) containing ill formed sentences with grammatical errors, typographical errors and abbreviations. The entry numbered 2, has detailed information of the steps taken to resolve the issue. The entry 4, has very limited information and hence is of little value. The characteristics of the textual information available in the maintenance of 4 IT applications is shown in Table 2. Textual data is small in terms of the number of words in a process instance log.

Table 1. Unstructured textual information captured during IT maintenance process

No.	Communication log of the problem tickets recorded by knowledge workers
1	emailed user. *waiting for user to get back to me* emailed user. looking for response User confirmed that the issue is not replicated. Hence closing the incident
2	Left a voicemail for customer at the number provided in this ticket Requested he call option (one) for further assistance **Validated userid in the portal, made in Synch** **Manually made in Synch with that of GUI** Call made both on office phone and cell *Voice sent on cell and office phone is not reachable* *2nd call made to the customer.* No response.. *3rd call made to the customer* No response. Call closed due to no prior response from the customer
3	**incorrect logon locks**. unlocked the ID and reset the password pinged user via IM *John confirmed to close the incident*
4	Received confirmation from user, closing the incident

Table 2. Characteristics of textual data in process logs of real-life IT application maintenance process

Application	Number of process instances	Number of sentences	Average number of words per sentence	Average number of words per process log
Application security	684	2235	10.25	44.35
Portal	210	1569	14.11	118.02
HR system	490	1482	11.87	41.38
Reporting	832	1267	9.71	20.02

However, these logs reflect some common situations that arise when performing an activity. For example, 'Unavailability of the customer' could be a situation or a task context, and could impact the time taken to perform the task. The log contains both, (i) information relevant to the specific process or task, and (ii) information that represents context. Hence, the textual data can refer to multiple topics. In the following section, we describe the background of concepts that can be applied to mine relevant information from the logs, specifically related to identifying multiple topics from textual documents.

3   Background

This section presents well known natural language processing techniques that can be used together to mine contextual information from process logs.

3.1   Notations

The textual information logged during the execution of a process instance can be considered as a text document. Let each document $d_i \in D$ represent textual information logged for respective process instance $p_i \in P$. Each document could comprise information on activities being performed, the actions taken when performing the activity and the situation or conditions during the execution of the activities. Hence, document d_i comprises of one or more topics of the topic set $T = \{t_1, t_2 \ldots t_T\}$ with some topics representing the context of the process instance. The problem can be represented as a multi-label categorization of textual logs.

We further assume that each document d_i is represented by smaller constituents that relate to one or more topics. The smaller constituents or chunks of text are called *segments*, which in turn contain one or more sentences. A segment is small enough to contain information relevant to a single topic. We believe that, in general, this assumption holds for communication logs containing short descriptions. Hence let S_i be the set of segments of document d_i, then $S = \bigcup_{i=1}^{|D|} S_i$, is a set of all segments. The goal is to find the topics T over S, and further find the topics for each document $T_i \subseteq T$ based on topics of the segments S_i of the document d_i, and hence the process instance p_i.

3.2   Segmenting Document

The goal of breaking down the document into segments, is to identify smaller constituents that represent distinct information related to tasks or their context. There are multiple ways of segmenting text. The suitability of the method is based on the characteristics of the textual information in the process logs.

1. *Phrase extraction* using parts-of-speech (POS) patterns has been used to extract text segments [6,21]. These are similar to regular expression patterns based on parts of speech. While, pattern based extraction has a high precision in extracting information, it has low recall as it filters phrases that do not match the POS pattern. For example, the phrases 're-provisioning completed', 'has been re-provisioned' and 're-provisioned and sent confirmation', have the same information, and yet have different POS tag patterns: 'VBG VBN', 'VBZ VBN VBN', 'VBN CC VBN NN' respectively (VBN is verb, CC is conjunction, and NN is noun, based on the listing of POS tags by Penn Treebank Project [18]). This method of segmentation is suitable when information logged by process participants is based on standardized templates.
2. *Parse Tree* is a rooted tree that represents the syntactic structure of a sentence based on a grammar. There are two ways of constructing parse trees: (1) constituency relation that is based on phrase structure grammar, (2) dependency relation that is based on relations among words. Constituency parser can be used to break down the sentence to extract smaller noun or verb phrases. Noun and verb phrases can be used as segments of the document. Parse trees are suitable when there is very sparse data reported by the process participants. In such scenarios the information extracted, is limited to key actions recorded during process execution. For example, from the communication log on the first row in Table 1, verb phrases such as 'emailed user', 'waiting for user', 'looking for response' can be extracted by using constituency parser.
3. *Extractive summarization* is an automatic text summarization method that, produces a summary of the text while retaining key information in a document [2]. There are two well known methods to summarization (i) abstractive summarization, and (ii) extractive summarization. Extractive summarization identifies important sections of the text and generates them verbatim. Distinct sentences of the document summary can be used as segments. Summarizing text is suitable when there verbose comments logged by process participants.

3.3   Clustering Methods

The extracted text segments can be categorized and grouped using different clustering methods. We briefly discuss common clustering methods and their suitability to grouping textual data available in process logs:

1. *Topic Modeling* Clustering approaches such as latent semantic analysis [20], probabilistic latent semantic analysis (pLSA) [14] and latent Dirichlet allocation (LDA) [3] have been used to identify representative set of words or topics. These approaches identify topics by exploiting the co-occurrence of

words within documents and are well suited for multi-topic text labeling. However, they are not suitable for short documents containing limited number of words and sentences. Hence, while these methods are widely used in multi-class text categorization, they are unsuitable for textual data available in process logs.

2. *Partition based clustering* such as k-Means, k-Mediods, are the most widely used class of clustering algorithms [13]. These algorithms form clusters of data points, by iteratively minimizing a clustering criterion and relocating data points between clusters until a (locally) optimal partition is attained. An important requirement of partition based methods is the number of partitions or K as input.

3. *Affinity Propagation* is one of the recent state-of-the-art clustering methods that has better clustering performance than partition based approaches such as k-Means [7]. Affinity propagation identifies a set of 'exemplars' and forms clusters around these exemplars. An exemplar is a data point that represents itself and some other data points. The input to the algorithm is pair-wise similarities of data points. Given the similarity matrix, affinity propagation starts by considering all data points as exemplars and runs through multiple iterations to maximize the similarity between the exemplar and their member data points.

3.4   Text Similarity

Next, we focus on the key aspect of any clustering algorithm; the choice of (dis)similarity function or distance metric between data points (text segment pairs). A text segment, is represented as a vector and distance functions such as Euclidean distance or similarity functions such as cosine similarity are used.

1. *Bag-of-Words (BOW)*: Each text segment is represented as vector of word counts of dimensionality $|W|$, where W is the entire vocabulary of words.

2. *TF-IDF*: The bag-of-words representation divided by each word's document frequency (number of text segment it occurs). The representation ensures that commonly occurring words are given lower weight.

3. *Neural Bag-of-Words (NBOW)*: Each text segment is represented as a mean of the embeddings of words contained in the text segment. The embeddings of words are obtained using the word2vec tool [19]. As the word vectors retain the semantic relationships, the distances between embedded word vectors can be assumed to have semantic meaning.

4. *Word mover distance (WMD)*: WMD is suitable for short text documents (or text segments). It uses word2vec embeddings [16]. The word travel cost (or euclidean distance), between individual word pairs is used to compute document distance metric. The distance between the two documents is the minimum (weighted) cumulative cost required to move all words from d_i to d_j. When there are documents with different numbers of words, the distance function moves words to multiple similar words.

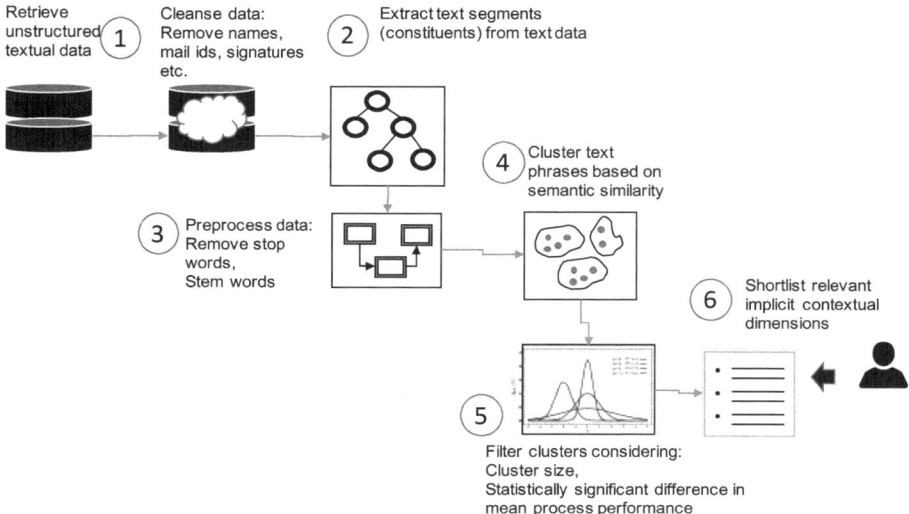

Fig. 1. Overall approach to identify implicit contextual dimensions

4   Overall Approach

Our approach to infer or identify implicit context is organized into multiple steps, as shown in Fig. 1. The approach comes down to answering three key questions: (i) What are the common situations and actions taken by the performers of a process during its execution? (ii) How many process instances are related to these situations? - is this a common or a rare situation? and (iii) Are these representative of process context and do they impact the performance outcome of the process? The steps of the approach are discussed in detail:

4.1   Text Retrieval and Cleansing

A tuple $\langle pid, ppi, text\_data \rangle$ containing the process instance identifier (pid), the process performance indicator (ppi) [4], and the unstructured textual information is extracted from execution logs. The use of each of these attributes, will be described in the following steps. The $text\_data$ for each process instance is referred to as a document. The document is processed to remove the names of people, IP addresses, HTTP addresses, and other textual data such as email signatures, phone numbers, that would not represent common actions or situations. The cleansing uses named entity recognizer[1], to detect person names, organization names. IP addresses, phone numbers, email addresses are cleaned from the text using regular expression parsers.

---

[1] https://nlp.stanford.edu/software/CRF-NER.html.

4.2   Text Segmentation

In this step the document is broken down into text segments by extracting summaries, or by extracting phrases using constituency parsing. A suitable method is chosen based on the characteristics of textual log (sparsity, verbosity, or variety), as described in Sect. 3.2. Hence we have $\langle pid, text\_segment \rangle$.

4.3   Text Preprocessing

Each text segment goes through standard preprocessing steps (i) lemmatization, where the base form of the words in the text segment are derived (e.g. - allocate, allocation, allocating are replaced by their lemma 'allocate'). (ii) Stop word removal, where very frequent words that are likely to appear in all the documents and contain little information, are removed.

4.4   Clustering

The text segments are clustered using one of the similarity measures described in Sect. 3.4. This step results in grouping process instances having similar text segments. The process instance associated to each text segment and its performance indicator is used to form a tuple $\langle pid, cluster\_id, text\_segment, ppi \rangle$.

4.5   Filtering Clusters

The goal of this step is to identify clusters of text segments, that are important and useful to a domain expert and help discern contextual dimensions. Two filters can be applied:

Size Filter: The number of process instances associated with a cluster is a good indicator of its importance. Intuitively, if the size is very large, then the information content is a part of normal execution of the task. For example, if the number of process instances associated to the phrase 'confirming and closing loan application' is very large, it is indicative of a normal procedure. Similarly, a cluster containing very few process instances may not be useful as it may indicate an exception and has to be handled as a part of the process exception or process error management. An upper and lower bound on number of process instances is set to filter clusters.

Process Performance Filter: This filter helps identify clusters that have an impact on the performance indicators of the process. The performance indicators of a process can be the completion time, the quality outcome of the process, or any other process indicator as detailed in [4]. To verify if the performance indicators of the process instances of a cluster are significantly different from other process instances, we consider two sample groups - (i) cluster group, and (ii) other group. Performance indicators of all process instances in a cluster are taken as one sample (cluster group). Performance indicators of a randomly chosen set of process instances from other clusters are considered as the second independent sample (other group). The Mann-Whitney U test is used to compare statistically, the difference in the performance indicators of the two groups. The test is run with multiple random samples of *other group* to reduce false positives or Type 1 error. The Mann-Whitney U test is one of the powerful nonparametric tests that makes no assumption on the distribution of data and is relevant for groups with small sample sizes (as clusters could be containing 10 process instances).

4.6   Context Identification

The final step of the approach is a manual verification by domain experts on the filtered set of clusters. The description in the text segments of filtered clusters are used by the domain experts to identify contextual situations that impact the performance of the process.

5   Experimental Evaluation

We first evaluate and compare the segment based clustering using different clustering methods, and similarity measures, on a benchmark set of multi-topic documents, as there is no benchmark textual data of business process available to evaluate the approach. Next, the overall approach detailed in Sect. 4, is used on a real-life business process textual log to identify the clusters that indicate contextual information.

5.1   Evaluating Clustering of Text Segments

The Reuters-21578 text categorization collection is a text categorization benchmark [29]. The *Mode Apte* evaluation, is used in which unlabeled documents are removed. There are 10787 documents that belong to 90 categories. The collection has a training set containing 7768 documents and a test set containing 3019 documents. Two main constraints are set up on the data: (1) each document should be assigned to at least 3 topics or categories, (2) each category or topic must have at least 1% of the documents. The training set is used to set the parameters for affinity propagation and choosing K for k-Means, and group text segments into the same number of clusters as the categories in the collection (68 categories in our case).

The quality of segment based clustering is evaluated on the test data containing over 900 segments on 95 multi-labeled documents, using the commonly used criterion of *precision, recall* and *F1 measure* [27]. Two approaches are used to compute the measures for multiple categories. The Precision, Recall, F1-measure is computed for each category. Finally, the overall measure is obtained by averaging category specific Precision, Recall and F1 measure. This is known as macro-averaging ($Prec_M, Rec_M, F1_M$). The other approach is based on computing a confusion matrix of all the categories by summing the documents that fall in each of the four conditioned sets, namely true positives, true negatives, false positives, and false negatives. The Precision, Recall and F1 measure is computed with the overall confusion matrix. This second measure is known as micro-averaging ($Prec_\mu, Rec_\mu, F1_\mu$).

The results are presented in Table 3. Text segments for each document are created by using extractive summaries. As K-Means algorithm is based on euclidean distance between two pairs, word mover distance is not evaluated. The results indicate that using affinity propagation based clustering, provides better F1 scores as compared to K-Means. Euclidean distance of NBOW and WMD measures result in higher macro-average and micro-average F1.

Table 3. Comparative evaluation of multi-class categorization for various distance measures and clustering methods

Clustering	Similarity	Macro-average			Micro-average		
		$Prec_M$	Rec_M	$F1_M$	$Prec_\mu$	μ	$F1_\mu$
K-Means	BOW	0.772	0.442	0.491	0.385	0.490	0.431
	TF-IDF	0.583	0.586	0.534	0.552	0.447	0.495
	NBOW	0.665	0.538	0.530	0.55	0.467	0.503
Affinity propagation	BOW	0.705	0.450	0.448	0.341	0.535	0.417
	TD-IDF	0.648	0.548	0.568	0.614	0.483	0.541
	NBOW	0.637	0.626	**0.580**	0.570	0.516	**0.542**
	WMD	0.652	0.593	**0.584**	0.631	0.470	**0.540**

5.2   Context Mining from Text Logs

The overall approach of identifying contextual information is evaluated on an IT maintenance process of 3 different applications of a large media and entertainment organization. The textual data recorded varies significantly for different application domains such as security, human resources, finance and web portal. The process consists of four main tasks: (1) customer creates an application problem ticket, (2) the worker acknowledges the receipt the ticket, (3) the worker analyzes the issue and resolves the problem, (4) on resolving the problem, the worker confirms with the user, and (5) the worker closes the ticket. At each step, the workers log their findings or progress. In some cases, emails sent or received by the customer and the worker is logged in the system. We analyze the communication or task logs associated with each process instance.

To evaluate the overall approach of mining contextual factors from textual data, the pipeline of steps detailed in Sect. 4 is executed. Table 4 presents the descriptions derived from the text segments in the filtered clusters. As shown, for the 'Security' application, of the 2493 text segments extracted from all the process instance documents, clustering using affinity propagation with WMD, results in 119 groups or categories. The mean completion times of the process instances in these groups is compared to mean completion time of a random number of other process instances. A statistical significance in the mean completion time (the performance outcome), is used to filter few clusters. Further, a filtering of clusters is done based on the size of the cluster. For example, 'confirm and close incident' is a very common text segment that is identified and associated with several process instances. It occurs in 50% of the process instances. It may hence, be a process completion step and not a situation or context. The highlighted descriptions in the table are examples of context.

Based on the cluster labels in Table 4 (that are derived from common text in the clusters), for the security application, it is observed that any process instance associated with *reset password* has lower completion time (indicated with a + sign in the table), as the task is extremely specific. The clusters further highlight a key situation of not being able to contact the customers, leading to the process being set to 'pending' status and the completion time being much higher than other process instances. Identifying such a situation can help re-design the process to account for customer unavailability. Similarly in the maintenance of the portal application, waiting for *more information* from the user leads to higher completion time of such tasks. A template with all relevant information recorded by the customers when creating the problem ticket, could be a plausible solution. In the HR domain application, the number filtered clustered were limited and the clusters did not provide useful insights on context.

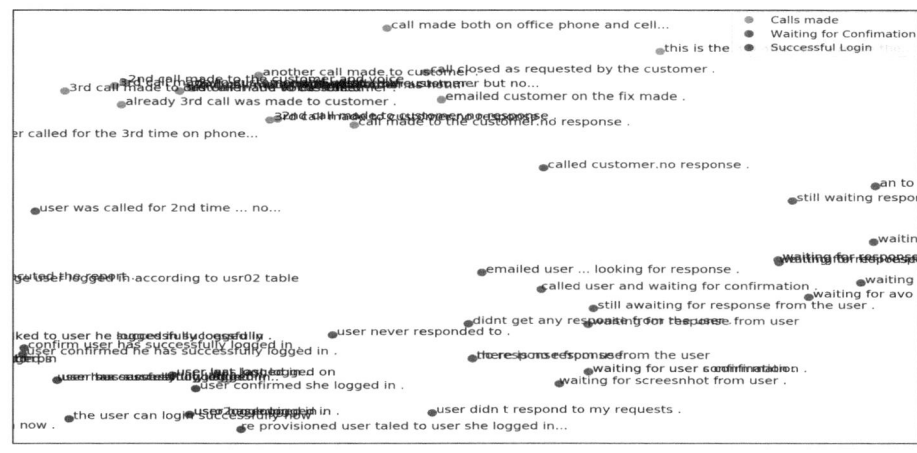

Fig. 2. Visualization of a subset of clusters

Figure 2 visually depicts a subset of clusters of the textual segments. The NBOW vectors of text segments is represented on a two dimensional space. The textual segments are the noun and verb phrases extracted using constituency parser.

Table 4. Filtered clusters of IT application maintenance process logs, (+) indicates clusters has lower completion times

App. Domain	# Text Segments	#Clusters	# Filtered	Cluster labels
Security	2493	119	13	1. **(2nd call, 3rd call) made to the customer**
				2. (researching, working, fixing) issue
				3. (asked, sent, mailed) to check again
				4. **waiting for (approval, confirmation)**
				5. could not (read, get, contact) user
				6. waiting for user
				7. **reset password (sent, mailed) user** (+)
				8. **changing status to pending**
				9. **tried calling the user**
				10. . . . . . .
Portal	2025	170	22	1. **sent to the user for (confirmation, information)**
				2. **waiting for user (confirmation, email)**
				3. **moved support issue to development**
				4. **getting more details on the issue**
				5. called and left a voice mail
				6. . . . . . .
HR system	2092	189	27	1. **(were, tied to, failed) data issues**
				2. closing the incident (+)
				3. **need to upgrade to breakfix**
				4. (write, call) back to me
				5. . . . . . .

5.3   Threats to Validity

Threats to *external validity* concerns the generalization of the results from our study. We have tried to limit this threat by evaluating it on textual data of 4 application domains, with over 300 users logging comments on over 2000 process instances. While insights can be drawn from our study, we do not claim that these results can be generalized in all business processes. However, the results serve as the basis of using textual data to discern relevant process context. Threats to *internal validity* arise when there are errors or biases. In our study, we have used standard implementations of distance functions and cluster analysis. The clustering and filtering approach required some configuration parameters such as the minimum and maximum size of the clusters. These should not impact the applicability of the approach. The choice of measurements is considered as a threat to *construct validity*. Appropriate measures such as precision and recall were not used on textual data in process logs due to non-availability of labeled data. However, we evaluated metrics on a multi-labeled benchmark data set to compare various methods of grouping textual information used in our study.

6 Related Work

The business process management community has experimented with use of unstructured textual data for various use cases:

Generating business process model from textual documents has been studied in some of the earlier work. Ghose et al. [12] propose a Rapid Business Process Discovery (R-BPD) framework and toolkit that employs text-to-model translation. Templates of commonly occurring textual cues or patterns are used to derive processes or task descriptions. Information extraction based approach is used to identify verb and noun phrases. In addition, recent work by Friedrich et al. presents an automatic approach of generating BPMN models from natural language text [8]. Sentence level analysis is done to extract performers and actions. This is followed by text level analysis where the relationships between sentences is used to determine links between actions and the control flow.

Teinemaa et al. exploit both unstructured text and structured attributes of cases for predictive business process monitoring [28]. The authors present a framework that extracts features from textual documents and evaluate different combinations of text mining and classification techniques to label executions as positive or negative.

There have been several efforts on using unstructured textual information available in problem tickets raised during IT application or service maintenance. There are approaches that use supervised learning to identify the right team or service agents for efficient ticket assignment [1,24]. Automatic recommendation of resolution for problem ticket based on similar nearest neighbors has been studied [30]. The underlying approach evaluates semantically similar past problem tickets and recommends appropriate resolution. Automatically analyzing natural language text in network trouble tickets has been studied by Potharaju et al. [21]. The authors present Netseive, a tool that infers problem symptoms, troubleshooting activities and resolution actions. Mani et al. [17] use clustering techniques and assign salient labels to group similar problem tickets. They use a combination of Lingo, a phrase based clustering method and N-gram extraction to identify phrases or cluster labels. However, they do not evaluate the clusters and their performance outcomes. In this work, we use an IT service management process for our study and evaluate different segment based clustering methods. Our approach further evaluates the clusters and analyzes the performance of process instances in these clusters.

Context-aware business process modeling has focused on design and specification of contextual attributes or dimensions [22,23]. There have been efforts on designing and evaluating impact of context on the process performance [10,11], and task allocation decisions [25,26].

Kiseleva et al. [15] introduced the notion of implicit and explicit context for predicting user behavior in eCommerce applications. The web user's age, gender and other known attributes are considered as explicit context, while information such as the purchase intent of the user is not known and is considered to be hidden context.

We propose a method of using textual information available in the process execution logs to uncover contextual dimensions.

7   Conclusion and Future Work

In this study, we proposed a novel approach of leveraging textual logs captured during a process execution for identifying useful and relevant situations or context. Using unstructured information extraction methods, we developed our approach of clustering process instances or tasks into unified groups, correlating them with process outcome and identifying a subset of situations that are correlated to the performance outcome. Our approach is quite general, and can be applied to different application domains. In future, we intend to explore filtering approaches beyond cluster size and performance outcomes. We also want to explore possibilities of automating identification of contextual situations by using labeled dataset and supervised learning techniques.

References

1. Agarwal, S., Sindhgatta, R., Sengupta, B.: SmartDispatch: enabling efficient ticket dispatch in an IT service environment. In: KDD, pp. 1393–1401 (2012)
2. Allahyari, M., et al.: Text summarization techniques: a brief survey. In: CoRR abs/1707.02268 (2017)
3. Blei, D.M., Ng, A.Y., Jordan, M.I.: Latent Dirichlet allocation. J. Mach. Learn. Res. **3**, 993–1022 (2003)
4. del-Río-Ortega, A., Resinas Arias de Reyna, M., Durán Toro, A., Ruiz-Cortés, A.: Defining process performance indicators by using templates and patterns. In: Barros, A., Gal, A., Kindler, E. (eds.) BPM 2012. LNCS, vol. 7481, pp. 223–228. Springer, Heidelberg (2012). https://doi.org/10.1007/978-3-642-32885-5_18
5. Dourish, P.: What we talk about when we talk about context. Pers. Ubiquitous Comput. **8**(1), 19–30 (2004). ISSN 1617-4909
6. Fader, A., Soderland, S., Etzioni, O.: Identifying relations for open information extraction. In: Proceedings of the 2011 Conference on Empirical Methods in Natural Language Processing, pp. 1535–1545 (2011)
7. Frey, B.J., Dueck, D.: Clustering by passing messages between data points. Science **315**(5814), 972–976 (2007). https://doi.org/10.1126/science.1136800
8. Friedrich, F., Mendling, J., Puhlmann, F.: Process model generation from natural language text. In: Mouratidis, H., Rolland, C. (eds.) CAiSE 2011. LNCS, vol. 6741, pp. 482–496. Springer, Heidelberg (2011). https://doi.org/10.1007/978-3-642-21640-4_36
9. Ghattas, J., Soffer, P., Peleg, M.: A formal model for process context learning. In: Rinderle-Ma, S., Sadiq, S., Leymann, F. (eds.) BPM 2009. LNBIP, vol. 43, pp. 140–157. Springer, Heidelberg (2010). https://doi.org/10.1007/978-3-642-12186-9_14
10. Ghattas, J., Soffer, P., Peleg, M.: Improving business process decision making based on past experience. Decis. Support Syst. **59**, 93–107 (2014)

11. Ghattas, J., Peleg, M., Soffer, P., Denekamp, Y.: Learning the context of a clinical process. In: Rinderle-Ma, S., Sadiq, S., Leymann, F. (eds.) BPM 2009. LNBIP, vol. 43, pp. 545–556. Springer, Heidelberg (2010). https://doi.org/10.1007/978-3-642-12186-9_53

12. Ghose, A., Koliadis, G., Chueng, A.: Process discovery from model and text artefacts. In: 2007 IEEE International Conference on Services Computing - Workshops (SCW 2007), 9–13 July 2007, Salt Lake City, Utah, USA, pp. 167–174 (2007)

13. Hartigan, J.A., Wong, M.A.: Algorithm as 136: a K-Means clustering algorithm. J. R. Stat. Soc. Ser. C (Appl. Stat.) **28**(1), 100–108 (1979). ISSN 00359254, 14679876

14. Hofmann, T.: Probabilistic latent semantic indexing. In: Proceedings of the 22nd Annual International ACM SIGIR, SIGIR 1999, 15–19 August 1999, Berkeley, CA, USA, pp. 50–57 (1999)

15. Kiseleva, J.: Context mining and integration into predictive web analytics. In: 22nd International World Wide Web Conference, WWW 2013, Rio de Janeiro, Brazil, 13–17 May 2013, Companion Volume, pp. 383–388 (2013)

16. Kusner, M.J., et al.: From word embeddings to document distances. In: Proceedings of the 32nd International Conference on Machine Learning, ICML 2015, Lille, France, 6–11 July 2015, pp. 957–966 (2015)

17. Mani, S., et al.: Panning requirement nuggets in stream of software maintenance tickets. In: Proceedings of the 22nd ACM SIGSOFT International Symposium on Foundations of Software Engineering, (FSE-22), Hong Kong, China, 16–22 November 2014, pp. 678–688 (2014)

18. Marcus, M., et al.: The Penn Treebank: annotating predicate argument structure. In: Proceedings of the Workshop on Human Language Technology, HLT 1994, pp. 114–119. Association for Computational Linguistics, Plainsboro (1994). ISBN 1-55860-357-3

19. Mikolov, T., et al.: Efficient estimation of word representations in vector space. In: CoRR abs/1301.3781 (2013)

20. Osiński, S., Stefanowski, J., Weiss, D.: Lingo: search results clustering algorithm based on singular value decomposition. In: Kłopotek, M.A., Wierzchoń, S.T., Trojanowski, K. (eds.) Intelligent Information Processing and Web Mining. AINSC, vol. 25, pp. 359–368. Springer, Heidelberg (2004). https://doi.org/10.1007/978-3-540-39985-8_37

21. Potharaju, R., Jain, N., Nita-Rotaru, C.: Juggling the Jigsaw: towards automated problem inference from network trouble tickets. In: Proceedings of the 10th USENIX Symposium on Networked Systems Design and Implementation, NSDI 2013, Lombard, IL, USA, 2–5 April 2013, pp. 127–141 (2013)

22. Saidani, O., Nurcan, S.: Context-awareness for adequate business process modelling. In: Proceedings of the Third IEEE International Conference on Research Challenges in Information Science, RCIS 2009, Fès, Morocco, 22–24 April 2009, pp. 177–186 (2009)

23. Saidani, O., Rolland, C., Nurcan, S.: Towards a generic context model for BPM. In: 48th Hawaii International Conference on System Sciences, HICSS 2015, Kauai, Hawaii, USA, 5–8 January 2015, pp. 4120–4129 (2015)

24. Shao, Q., et al.: Efficient ticket routing by resolution sequence mining. In: Proceedings of the 14th ACM International Conference on Knowledge Discovery and Data Mining, KDD 2008, Las Vegas, Nevada, USA, pp. 605–613 (2008). ISBN 978-1-60558-193-4

25. Sindhgatta, R., Ghose, A., Dam, H.K.: Context-aware analysis of past process executions to aid resource allocation decisions. In: Nurcan, S., Soffer, P., Bajec, M., Eder, J. (eds.) CAiSE 2016. LNCS, vol. 9694, pp. 575–589. Springer, Cham (2016). https://doi.org/10.1007/978-3-319-39696-5_35

26. Sindhgatta, R., Ghose, A., Dam, H.K.: Context-aware recommendation of task allocations in service systems. In: Sheng, Q.Z., Stroulia, E., Tata, S., Bhiri, S. (eds.) ICSOC 2016. LNCS, vol. 9936, pp. 402–416. Springer, Cham (2016). https://doi.org/10.1007/978-3-319-46295-0_25

27. Sokolova, M., Lapalme, G.: A systematic analysis of performance measures for classification tasks. Inf. Process. Manag. **45**(4), 427–437 (2009)

28. Teinemaa, I., Dumas, M., Maggi, F.M., Di Francescomarino, C.: Predictive business process monitoring with structured and unstructured data. In: La Rosa, M., Loos, P., Pastor, O. (eds.) BPM 2016. LNCS, vol. 9850, pp. 401–417. Springer, Cham (2016). https://doi.org/10.1007/978-3-319-45348-4_23

29. Yang, Y., Liu, X.: A re-examination of text categorization methods. In: Proceedings of the 22nd Annual International ACM SIGIR, SIGIR 1999, Berkeley, California, USA, pp. 42–49 (1999)

30. Zhou, W., et al.: Resolution recommendation for event tickets in service management. IEEE Trans. Netw. Serv. Manag. **13**(4), 954–967 (2016). ISSN 1932–4537

Advanced Simulation of Resource Constructs in Business Process Models

Sander P. F. Peters[✉], Remco M. Dijkman, and Paul W. P. J. Grefen

Eindhoven University of Technology, Eindhoven, The Netherlands
{s.p.f.peters,r.m.dijkman,p.w.p.j.grefen}@tue.nl

Abstract. Simulation is often used as a tool to assess the performance of business processes. However, current business process simulation engines do not support advanced resource constructs, such as work allocation strategies and case attributes. Using only basic resource constructs leads to performance metrics that deviate significantly from the real process performance. Therefore, a clear need arises for simulation engines that incorporates advanced resource constructs. Addressing this need, we present the resource patterns that should be supported by simulation engines, a conceptual model to support them, and a prototype implementation of this conceptual model. The model and engine are evaluated in a simulation experiment that highlights utilization rates under different conditions. This experiment shows that the advanced resource constructs significantly outperform the basic resource constructs. From this we can also conclude that existing simulation engines must be extended with advanced resource constructs to properly simulate processes from practice that use these constructs.

1 Introduction

Analysis of the performance and feasibility of business processes is important to evaluate the effects of business process reengineering and redesign efforts [6]. In order to assess the performance of these business process models, simulation is often used. However, Recker [15] shows that more research into the simulation of business process models is needed.

First, business process simulation engines should take into account basic simulation parameters, as described in the standard work on simulation by Law and Kelton [12]: each simulation engine should support a warm-up period, replications and confidence intervals. A warm-up period ensures correct simulation results, since the empty system at the beginning of the simulation can pollute the final performance of the model. In order to obtain confidence intervals using the central limit theorem, replications need to be in place in the simulation engine.

Second, a strong resource perspective is needed to ensure that the process models correctly represent reality. However, we will show in Sect. 2 that commonly used business process simulation engines have problems coping with advanced resource constructs, such as resource dependencies and queueing strategies. We focus on simulation engines that use the Business Process Model

© Springer Nature Switzerland AG 2018
M. Weske et al. (Eds.): BPM Forum 2018, LNBIP 329, pp. 159–175, 2018.
https://doi.org/10.1007/978-3-319-98651-7_10

and Notation (BPMN) [14], because this has become the de-facto standard for modeling business processes. BPMN mainly focuses on the control-flow perspective of a process, while the resource perspective in the BPMN language is limited to the lanes and pools concepts [19]. In contrast, business process execution languages like YAWL [20] have extensive support for the resource perspective, which indicates that there is a need for supporting the resource perspective. The advanced resource patterns used in these execution languages are also defined in Russell et al. [18].

The consequence of the shortfalls of BPMN simulation engines, is that the results obtained by them will be less valid for the real-life process that is simulated. In our evaluation we will indeed show that BPMN models, in which advanced resource patterns are not included, have simulation results that deviate significantly from models, in which they *are* included. Consequently, we can conclude that BPMN models cannot be used to draw valid conclusions for business processes from practice that use advanced resource constructs, such as the case handling principle.

Therefore, the aim and contribution of this paper is to present the conceptual model and behavior of a simulation engine that supports the advanced simulation requirements outlined above. We also present a prototype implementation of this engine. This contributes to laying the foundation for the simulation of complex business processes with the complete inclusion of resources for real world processes, which can enhance business process engineering and redesign efforts.

Against this background, the remainder of this paper is structured as follows. Section 2 describes the state of the art of existing business process simulation engines. In Sect. 3 an extension of process models is proposed to make them fit for simulation purposes, solving the limitations of the existing models. In Sect. 4 the formal behavior of the new simulation engine is provided by means of a state-space and state-space transitions. Section 5 describes the implementation of the simulation engine from Sect. 4 and the evaluation of the engine. The focus in this section is on the effect of resource dependencies on the utilization of resources. In Sect. 6 the conclusion based upon the evaluation of the new simulation model is presented and future work is discussed.

2   Related Work

Table 1 provides the overview of the support of existing simulation engines for advanced simulation concepts. In order to assess the selected simulation engines, several criteria are taken from literature. These criteria focus on the support a simulation engine provides for the simulation of business processes with a special focus on the resource perspective. For each feature available in a tool there is a + symbol. If a feature is not available in a tool a – symbol is used. For cases where a feature can be found, but only in the form of a complex workaround or in a limited manner, there is a ± symbol.

First basic flow criteria are assessed, based upon the paper of Van der Aalst et al. [24], three basic criteria are selected: sequence of activities, parallel execution of activities and branching or choice in activities. These have been chosen

Table 1. Analysis of simulation engines

	Tool:	Arena	CPN	Adonis	BIMP	Bizagi	BPSim	IBM	Signavio	TIBCO	Prototype
Basic Flow	Sequence	+	+	+	+	+	+	+	+	+	+
	Parallelism	+	+	+	+	+	+	+	+	+	+
	Branching	+	+	+	+	+	+	+	+	+	+
Activities	Arrival Distributions	+	+	+/-	+	+	-	+	-	+	+
	Duration Distributions	+	+	-	+	+	-	+	+/-	+	+
	Resource Requirements	+	+	+	+	+	+	+	+	+	+
	Cost per Activity	+	+	+	+	+	+	+	+	+	+
Resources	Capacity	+	+	+	+	+	+	+	+	+	+
	Roles	+	+	+	+	+	+	+	+	+	+
	Schedules	+	+	+	+	+	+	+	+	+	+
	Cost of Usage	+	+	+	+	+	+	+	+	+	+
	Multiple Roles	-	+/-	+	-	-	-	+	-	+	+
Advanced Constructs	Queueing Strategies	+	+/-	-	-	-	-	-	-	-	+
	Separation of Duties	+/-	+/-	-	-	-	-	-	-	+	+
	Case Handling	+/-	+/-	-	-	-	-	-	-	-	+
	Allocation Strategies	-	+/-	-	-	-	-	-	-	-	+
	Case Attributes	+	+	-	-	-	-	-	-	+	+
Simulation	Duration	+	+	+	+	+	+	+	+	+	+
	Warm-Up Period	+	+	-	-	-	-	+	-	-	+
	Replications	+	+	+	-	+	-	-	-	-	+
	Confidence Intervals	+	+	-	-	-	-	-	-	-	+

since these are in the top 10 of most used constructs [29]. Next to these basic flow criteria, the paper of Tumay [22] indicates that also activities and resources are basic elements in a simulation. Tumay [22] shows that each simulation needs an arrival process, also each activity can be associated with a duration and a cost function. The duration of an activity and arrival rate can consist of mathematical distributions. Activities can also require resources to be executed in the process model. Resources, according to the paper of Russell et al. [18], should be modeled having a certain capacity, role and schedule. Also resources may have costs of usage associated with them. Therefore all simulation engines are assessed on these criteria for activities and resources.

The paper of Russell et al. [18] also defines several different allocation strategies for resources, in this survey the focus will be on the constructs of separation of duties and case handling from the patterns defined. According to Russell et al. [18] separation of duties is described by as: "*The ability to specify that two tasks must be allocated to different resources in a given workflow case*" and case handling is described as: "*The ability to allocate the work items within a given workflow case to the same resource*". The paper of Tumay [22] indicates that queueing strategies are important for the way entities are sequenced in queues at activities. Furthermore allocation strategies and case attributes are analyzed as important aspects in the simulation [18]. Next to these specific business process simulation properties, general properties of simulation engines will be reviewed in terms of duration of the simulation, ability to incorporate warm-up time, use of replications and the reporting of statistics making use of confidence intervals. These properties are prescribed by Van der Aalst et al. [23] as the main important properties to create a sound simulation experiment.

The simulation engines which are selected for this survey are composed of two types of tools: general purpose simulation tools and business process simulation tools, with the main focus on BPMN simulation engines. Based upon the paper from Jansen-Vullers and Netjes [7] two general purpose simulation tools are identified as Arena [9] and CPN Tools [8]. Furthermore all tools mentioned in the book Fundamentals of BPM [5] in Chap. 7.4.3 Simulation Tools will be assessed. Unfortunately ARIS, OpenText, Oracle and Savvion have no public available simulation tool to assess and therefore are excluded from the survey. Also the developed prototype as discussed in this paper will be scored on the criteria presented in Table 1.

As can be seen in Table 1 the two general purpose simulation tools perform quite well on mainly all aspects, only on the resource allocation rules they score a ±. In contrast the regular business process simulation tools perform worse on the resource allocation rules, where barely any advanced construct is supported. Furthermore there are major issues in general simulation properties, lack of confidence intervals, replications and warm-up periods makes most tools unfit for proper simulation according to the paper of Van der Aalst et al. [23]. General purpose simulation tools are less in favor than the business process simulation tools, which are easier to use and do not take a steep learning curve to model processes, since they use the BPMN modeling language [21, 27].

3   Advanced Simulation Model

The basic idea to simulating a business process model is to transform it into a queueing network [12]. In order to create that network, each activity in the BPMN model is transformed into a queue that contains cases on which that activity must be performed. Whenever the activity has been performed for a case in the queue, the case is put into the queue of the next activity that should be performed on it, conform the behavior that is described by the BPMN model.

In addition to this basic translation, which is common for all BPMN simulation engines, we propose constructs to simulate the following advanced constructs: case attributes, resource schedules, resources acting in multiple roles, queueing principles, resource dependencies, and simulation parameters. For illustration purposes, Fig. 1 shows a BPMN model that includes these advanced constructs. We explain these constructs in the following paragraphs.

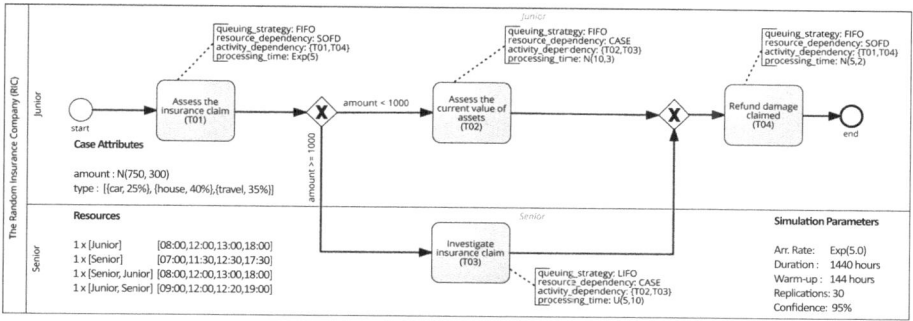

Fig. 1. BPMN Model with advanced parameters

3.1 Case Attributes

A case is an executable instance of a process, where each case can have certain attributes, also known as case data [28]. In practice, the choice between alternative paths through a process depends on the values of the case data. However, current simulation engines solve a choice between alternative paths by placing probabilities on the paths themselves. The benefit of associating alternative paths with the values of case data, is that this allows for interaction between multiple alternative paths to be captured.

To facilitate this, we define case attributes on the process level, where each attribute has a name and a value associated with it. Values for different attributes can be obtained from the probability distributions of the attributes as specified in the process model. Distributions can be either numerical or categorical, numerical distributions can be the normal distribution, the exponential distribution, the uniform distribution or a static value in this simulator. Categorical variables are uniformly distributed variables, where the attribute is assigned one of the values of the categorical variable. All values for the case are determined at the arrival of the case at the process. For example a case of an insurance claim process is of the category car (0.25), house (0.40) or travel (0.35) and has a certain amount of money claimed, which is normally distributed with an average of 1750 and a variance of 500.

3.2 Resource Schedules

According to Russell et al. [18] resources need to have a schedule. A schedule is an overview of the periods during which a resource is available for processing cases, also called the active state. When a resource is outside of the scheduled times, it is passive. During the time a resource is passive it will not actively participate in the execution of the business process. Each schedule is denoted by a chain of times in a day when a resource changes its state. For the first time in the schedule the resource becomes active and at each following event the state will change to passive if the resource was active and the state will change to active when the resource was passive. We will represent a schedule as a list of times at which the resource changes state from active to passive or vice versa, assuming that the resource starts passive, for example [08:00, 12:00, 13:00, 18:00]. Each resource will finish the last activity, if necessary in overtime, before becoming passive according to the schedule.

3.3   Resources Acting in Multiple Roles

According to Van der Aalst, Weske and Grünbauer [25] a resource can have multiple roles, based on the organizational model of the enterprise. To facilitate this, we define a resource as having a certain resource type, which is a list of roles belonging to a resource. A resource type can be a single role or a combination of multiple roles which exist in the same pool in a process model. Furthermore each of the resource types has an integer which represents the number of instances of the resource type. Each instance of a resource type is able to serve the roles which are associated with the type in the process and will stick to the order of the roles for prioritizing roles. For example if there are two roles, say A and B, resources can be of type [A], type [B], type [A,B] or type [B,A]. The quantities of resource types are defined as part of the process.

3.4   Queueing Principles

Queueing principles are the order in which a resource selects a case from a queue to handle it, this is also called the service discipline of the queue. The queue is the worklist which contains the work items offered to the resource. According to Righter et al. [16] queues can have different service disciplines, such as First In First Out (FIFO) and Last In First Out (LIFO). In the FIFO system the case first in the queue is also the first one to be handled by the resource. Where with the LIFO system the last case in the queue is handled first by the resource. Next to those two variants a random selection can be applied as a service discipline, in which the next case to be handled is selected randomly, or a priority system can be used to sort the queue according to the priority value of each case [12].

3.5   Resource Dependencies

Resource dependencies are specific dependencies on resource instances between activities in a process. From the paper of Russell et al. [18] two resource dependencies between activities can be identified: separation of duties (SOFD) and case handling (CASE). A resource dependency is an extra constraint on which resources may handle the activity. The separation of duties dependency yields that a different resource instances should execute the activities affected by the dependency. For the case handling dependency it is the other way around, it yields that the same resource instance should execute the activities affected by the dependency. Regularly each activity is not affected by any resource dependency and can be executed on its own by a resource instance. When there are resource dependencies between multiple activities for each activity in the dependency it should be denoted which type of dependency applies and which other activities are involved. For example if the activities A02 and A04 are subject to the resource dependency case handling both will have the resource dependency label CASE and the associated set of activities {A02, A04} in their properties denoted. If no resource dependency label is provided it is assumed that there is no resource dependency in the model for the given activity.

3.6   Simulation Parameters

In order to perform simulation, several main components should be defined in the BPMN model. According to Van der Aalst et al. [23] and Law and Kelton [12] the duration of the simulation, warm-up period, number of replications and confidence intervals are vital components of a simulation model. The warm-up period of the simulation is the amount of time where no performance metrics are recorded at the start of a simulation run [13]. This is needed because steady state simulation models start empty and idle, this empty state influences the performance metrics and therefore need to be corrected for. Replications are the repetition of a simulation with fixed inputs but different outputs due to different random number streams [3], these are needed to obtain reliable values for performance metrics. Using the values from the different replications confidence intervals can be computed using the central limit theorem [23]. These confidence intervals are a limited bandwidth where the actual value of a performance metric can be approached with a certain probability. These parameters should be set at the model level of a BPMN model in order to enable proper execution.

4   Advanced Simulation Behavior

In this section we formally define the behavior of a discrete event simulation that can simulate the concepts that are defined in the previous section. We define the behavior in terms of the statespace of a simulation model as it is specified in the previous section, the initial state of that model and the state transitions that the simulation model will take from the initial state.

The behavior is defined for the meta-model that is implicitly defined in the previous section and exemplified in Fig. 1. Due to space limitations, we only postulate the existence of the concepts that we need from the meta-model.

Definition 1 (Model Elements). *Let $\mathcal{I}$ be the set of all possible identifiers and $id : \mathcal{I}$ the function that returns a new unique identifier. Furthermore, let $\mathcal{L}$ be the set of all possible labels, $\mathcal{V}$ be the set of all possible data values, $\mathcal{S}$ be the set of all possible schedules and $\mathcal{D}$ be the set of all possible probability distributions.*

A simulation model defines a set of activities $\subseteq \mathcal{L}$ *and a set of* attributes $\subseteq \mathcal{L}$ *by their labels. It describes a set of* resourcetypes $\subseteq \mathcal{L} \times \mathbb{P}(\text{activities}) \times \mathcal{S} \times \mathbb{N}$ *(where $\mathbb{P}$ represents the powerset), such that each resource type $(l, as, s, n) \in$* resourcetypes *has a label l, a set of activities as that a resource of that type* can perform, a schedule s for the availability for resources of that type, and a number of resources n of that resource type. A schedule is a list of timestamps at which a resource changes state, a function to retrieve the next time of a state change is $t = getnext(s, currenttime)$, which returns the next time in a schedule at which the resource changes state. In addition a simulation model describes: a processing time distribution for each activity a, denoted as dist_a, an interarrival time distribution, denoted as dist_i, and a distribution for the values of each case attribute d, denoted as dist_d. We define the existence of a function sample : $\mathcal{D} \to \mathcal{V}$ that samples a value from a distribution.

Furthermore, a BPMN model describes a behavior that can be defined in terms of a token-game, as explained for example in Van Gorp and Dijkman [26] and Dijkman, Dumas and Ouyang [4]. Here, we assume the existence of functions that implement the token game. These functions should be implemented conform the BPMN semantics, as defined in these papers.

Definition 2 (Model Behavior). *Let $\mathcal{M}$ be the set of all possible markings in a token game on a BPMN model. Given a particular model, there is: a function* initialmarking : $\mathcal{M}$ *that returns the initial marking of the model for a starting case, a function* enabled : $\mathcal{M} \to \mathbb{P}(\text{activities})$ *that returns the set of activities that can happen in a given marking, and a function* fire : $\mathcal{M} \times \text{activities} \to \mathcal{M}$ *that defines the marking to which a model transitions when a particular activity happens in a particular marking.*

With these definitions, we can define the statespace of the simulation model. Since a discrete event simulation is based on a queueing network, we define an abstract 'queue' type.

Definition 3 (Queue). *Let queue and queue' be two queues and e be a potential element of a queue, we postulate a the existence of functions and constants:* [], *which represents the empty queue; queue'* = *put(queue, e), which places the element e in the queue, resulting in a new queue (the exact position at which the element e is placed in the queue may depend on the queueing policy, e.g. in a FIFO queue the element e is placed at the end, but in a LIFO queue the element e is placed at the start); e* $\in$ *queue, which returns true is the element e is in the queue; e* = *first(queue), which returns the first element in the queue; queue'* = *removefirst(queue), which removes the first element from the queue; and queue'* = *remove(e, queue), which removes element e from the queue.*

The parts of the statespace are: the queues that contain the cases that are queueing for a certain activity to be performed on it; the queues that contain the resources that are waiting to perform an activity that they can perform according to their resource type; a queue of simulation events that still have to occur in the simulation; and the current simulation time. Consequently, we define these elements as follows.

Definition 4 (Case, Resource, Event, Statespace). *Let $\mathcal{T}$ be the set of possible timestamps.*

*A **case** is a tuple (id, as), in which: $id \in \mathcal{I}$ is a unique identifier; as $\subseteq$ attributes $\times \mathcal{V}$ is a set that assigns values to attribute labels for the case.*

*A **resource** is a tuple (id, rt), in which: $id \in \mathcal{I}$ is a unique identifier; and $rt \subseteq \mathcal{L}$ is the label of the resource type of this resource.*

*A simulation **event** is a tuple $(t, type, i)$, in which: $t \in \mathcal{T}$ is the timestamp at which the event is scheduled to occur; type $\in$ {Arrival, Complete, Activate, Passivate} is the type of simulation event, which can either be the arrival of a new case in the system, the completion of an activity for a case, an activation of a resource or pacification of a resource; and i is the information that is associated with the event, which is a case if type = Arrival, a resource if type = Activate or Passivate, a tuple (r, c, a) if type = Complete, in which a is the activity that was just completed, r is the resource that was performing the activity and c is the case for which the activity was being performed.*

By convention, we refer to elements of a tuple by the labels that are defined for them above. E.g. for an event e, e_{type} refers to the type of e.

We can now define the statespace of a simulation model as follows.

Definition 5 (Statespace). *The statespace is of a simulation model is composed of:*

- *foreach activity a, a queue $queue_a$ of cases that are queueing for that activity;*
- *foreach resourcetype rt, a queue $queue_{rt}$ of resources that are queueing to perform activities;*
- *a queue $queue_s$ of simulation events ordered by their timestamp with the event with the lowest timestamp first in the queue;*
- *a function states : case $\rightarrow \mathcal{M}$ that associates cases with their current marking in the BPMN model;*
- *a set of assignments $\subseteq$ resource $\times$ case of resources currently working on a case;*
- *the currenttime of the simulation model; and*
- *a set passiveresources of resources that are passive.*

An example of a statespace is provided in Table 2. Six cases and five resources are active in the model. For activities where no resource is available, the cases are queued in the activity queues. For resources where all associated activity queues are empty are queued in the resource queues. The marking is not specifically defined, we illustrate the marking as activities marked with cases.

We can now define the algorithms that set the initial state of the simulation and that describe the behavior of the simulation. The initial state of the simulation is constructed through Algorithm 1. The algorithm creates: an empty queue of waiting cases for each activity; a queue of resources for each resource type, containing as many resources as specified by the resource type; the queue of simulation events, containing one new case arrival event and activation events

Table 2. Example of a statespace

Object	Value
Cases	$c_1 = (13,\{(\text{amount},1293),(\text{type,'car'})\})$, $c_2 = (14,\{(\text{amount},144),(\text{type,'travel'})\})$ $c_3 = (15,\{(\text{amount},1325),(\text{type,'house'})\})$, $c_4 = (16,\{(\text{amount},749),(\text{type,'car'})\})$ $c_5 = (17,\{(\text{amount},635),(\text{type,'house'})\})$, $c_6 = (18,\{(\text{amount},125),(\text{type,'travel'})\})$
Resources	$r_1 = (1, [\text{junior}])$, $r_2 = (2,[\text{senior}])$, $r_3 = (3,[\text{senior, junior}])$, $r_4 = (4,[\text{junior,senior}])$, $r_5 = (5,[\text{senior}])$
Activity queues	$queue_{T01}$: $[c_5,c_6]$; $queue_{T02}$: [ ]; $queue_{T03}$: [ ]; $queue_{T04}$:$[c_1]$
Resource queues	$queue_{jr}$: [ ]; $queue_{sr}$: $[r_2]$; $queue_{sr,jr}$: [ ]; $queue_{jr,sr}$: [ ]
$queue_s$	$[(00\!:\!12\!:\!45, \text{Arrival}, (19,\{(\text{amount},855),(\text{type,'travel'})\}))$, $(00\!:\!14\!:\!22,$ Complete, $(r_1,c_4,\text{T02}))$, $(00\!:\!15\!:\!37,$ Complete, $(r_3, c_2,\text{T04}))$, $(00\!:\!16\!:\!58,$ Complete, $(r_4,c_3,\text{T03}))$, $(00\!:\!17\!:\!00, \text{Passivate}, r_1)$, $(00\!:\!17\!:\!15, \text{Passivate}, r_2)$ $(00\!:\!17\!:\!45, \text{Passivate}, r_3)$, $(00\!:\!18\!:\!00, \text{Passivate}, r_4)$, $(00\!:\!36\!:\!00, \text{Activate}, r_5)]$
Marking	$\{(\text{T04},c_1), (\text{T04},c_2), (\text{T03},c_3), (\text{T02},c_4), (\text{T01},c_5), (\text{T01},c_6)\}$
Assignment	$\{(r_1,c_4), (r_3,c_2), (r_4,c_3)\}$
currenttime	00:11:39
passiveresources	$\{r_5\}$

Algorithm 1. Create the initial state of the simulation model

---

1: **for all** activity $a \in$ activities **do** $queue_a = [\,]$
2: $newcase = (id(), \{(a, sample(dist_a))|a \in attributes\})$
3: $states(newcase) = initialmarking$
4: $currenttime = 0$
5: $queue_s = put([\,], (sample(dist_i), Arrival, newcase))$
6: **for all** resourcetype $(rt, as, s, n) \in$ resourcetypes **do**
7: $\quad queue_s = put(queue_s, (getnext(s, currenttime), Activate, rt))$
8: $assignments = \emptyset$
9: $passiveresources = \emptyset$
10: **for all** resourcetype $(rt, as, s, n) \in$ resourcetypes **do**
11: $\quad queue_{rt} = [\,]$
12: $\quad$ **for** i = 1 to n **do** $passiveresources = passiveresources \cup \{(id(), rt)\}$

---

for all resources; an empty set of assignments; and a set of passive resources, which initially contains all resources.

After the initial state, the simulation events in the event queue are processed by the simulation engine. Algorithm 2 specifies how this works. The algorithm processes simulation events in the order in which they should occur. If the sim-

---

Algorithm 2. Simulating the model by processing simulation events

---
1: **while** currenttime $<$ endtime **do**
2:     $(t, type, i) = first(queue_s)$
3:     $queue_s = removefirst(queue_s)$
4:     $currenttime = t$
5:     **if** type $=$ Arrival **then**
6:         **for all** $a \in enabled(states(i))$ **do** $queue_a = put(queue_a, i)$
7:         $newcase = (id(), \{(a, sample(dist_a)) | a \in attributes\})$
8:         $states(newcase) = initialmarking$
9:         $queue_s = put(queue_s, (currenttime + sample(dist_i), Arrival, newcase))$
10:    **else if** type $=$ Complete **then**
11:        $assignments = assignments - \{(i_r, i_c)\}$
12:        $queue_{rt'} = put(queue_{rt'}, i_r)$, where $rt' = i_{r_{rt}}$
13:        $states(i_c) = fire(states(i_c), i_a)$
14:        **for all** $a \in enabled(states(i_c))$ **do** $queue_a = put(queue_a, i)$
15:    **else if** type $=$ Activate **then**
16:        $passiveresources = passiveresources - \{i\}$
17:        $queue_{rt'} = put(queue_{rt'}, i)$, where $rt' = i_{rt}$
18:        $queue_s = put(queue_s, (getnext(rt'_s, currenttime), Passivate, i))$
19:    **else if** type $=$ Passivate **then**
20:        **if** $(r, c) \in assignments$, such that $i = r$ **then**
21:            find $(t, Complete, (r', c', a)) \in queue_s$, for which $r' = r$ and $c' = c$
22:            $queue_s = put(queue_s, (t, Passivate, r))$
23:        **else**
24:            $queue_{rt'} = remove(queue_{rt'}, i)$, where $rt' = i_{rt}$
25:            $passiveresources = passiveresources \cup \{i\}$
26:            $queue_s = put(queue_s, (getnext(i_{rt_s}, currenttime), Activate, i))$
27:    **for all** $a \in activities$, for which there exists $c \in queue_a$ and a resource type $(rt, as, n) \in resourcetypes$, such that $queue_{rt} \neq []$ and $a \in as$, using a fair distribution of resources over activities **do**
28:        $r = first(queue_{rt})$
29:        $queue_a = removefirst(queue_a)$
30:        $assignments = assignments \cup \{(r, c)\}$
31:        $e = (currenttime + sample(dist_i), Complete, (r, c, a))$
32:        $queue_s = put(queue_s, e)$

---

ulation event is an arrival event, the case that arrives is put in the queues of the activities that are enabled in the current (initial) marking of the case. Subsequently, an event is generated for the next arrival. If the simulation event is a completion event, the resource processing the completed activity is released, the next marking of the case is computed, and the case is put in the queues of the activities that are enabled in that marking. If the simulation event is an activate event, the passive resource is activated and a new passivate event is scheduled. If the simulation event is a passivate event, first there is checked whether the resource which should become passive is currently processing an activity for a case, if so a new passivate event is scheduled at the completion time of the activity. If the resource is not currently processing an activity, it is removed from the resource queue and moved to the passiveresources and a new activate event is scheduled. Subsequently, it is determined if there are any activities on waiting cases that can now be performed, because there are resources available to do so. Activities must be processed in an order that is 'fair', for example longest-waiting-time-first or round-robin.

5 Evaluation

The advanced simulation engine that is defined in this manner, improves on current BPMN simulation tools. Consequently, the hypothesis of this paper is that the current BPMN simulation tools produce analysis results that significantly deviate from reality.

In order to test this hypothesis, a simulation experiment will be conducted. In this paper the focus will be on resource dependencies, which consist of the concepts of separation of duties and case handling. In future work we will also investigate the effect of different queueing principles, case data, and other advanced resource patterns.

5.1 Simulation Methodology

The evaluation of the effect of the different resource dependencies is performed using simulation. The simulation will be run on a newly built discrete event simulation engine[1] which is able to handle the new queue mapping and the resource dependencies. This new simulator is developed in Java and makes use of the Desmo-J library. The simulation engine supports the resource dependencies, queueing strategies, case attributes and replications, warm-up periods and confidence intervals for the performance measures. A more extensive example process is available in the GitHub location to test the simulator.

The simulation model that will be used to test the hypothesis consists of two sequential activities served by the same resource type to simulate the patterns described by the resource dependencies. This provides three scenarios, where in all scenarios the same resource type is used, but which resource instance can handle the case depending on the resource dependency. Both of the activities have an exponentially distributed processing time with $\mu = 2.0$. The exponentially distributed arrival rate of the system will vary from 0.01 to the number of resources divided by the inter-arrival rate of 4.0. The number of resources in the simulation will increase from 2 to 10 resources. The duration of the simulation is 1440 h and the warm-up period is 192 h, which is visually confirmed by observing stability of the utilization rate, and 30 replications are performed per scenario. In Fig. 2 the simulation model is shown in BPMN 2.0.

In order to evaluate the effect of these dependencies the utilization rate of the simulation model is plotted against the theoretical predicted utilization rate

---

[1] Downloadable from: https://github.com/rmdijkman/simulator.

Fig. 2. Model for simulation

of the model. From Kleinrock [10] the theoretical utilization rate for an M/M/c queue is equal to: $\rho = \frac{\lambda}{c \cdot \mu}$. Where the assumption is made that the arrival process and the processing time are exponentially distributed and that there are c resources available for this queue. The theoretical expected utilization rate is adapted to correct for the number of queues which need to be served, where q stands for the number of queues which are served by the specific resource type: $\rho = \frac{\lambda \cdot q}{c \cdot \mu}$. Using this metric the theoretical utilization rate is calculated using the provided simulation parameters, assuming equal processing time for all queues. In the described simulation situation this is the case, so the metric can be used.

5.2   Results

From the results obtained in the simulation it can be concluded that the hypothesis is true, without taking the resource dependencies into account the result can differ as much as about fifty percent on performance of the resources in the process model. First the results from the separation of duties resource dependency will be discussed, next the case handling resource dependency is discussed. In each figure the three different scenario's are plotted against each other, where the blue line is the scenario without resource dependencies, the red dashed line is the scenario with separation of duties and the orange dash-dot line is the case handling scenario. For each line the 95% confidence interval is calculated and the differences are significant between the lines.

For the case where there are no resource dependencies it is observed that the achieved utilization in the simulation aligns perfectly with the theoretical utilization. This supports the claim that the simulation tool properly models the utilization of the resources.

The separation of duties resource dependency has a major impact in the scenario with only two resources as can be seen in Fig. 3. From a theoretical utilization of 0.9 there is a deviation where the resource is less optimally used in terms of utilization. For the scenario's with more resources the deviations between the separation of duties dependency and the situation without any resource dependencies are not significantly different. This can be explained by the fact that the restricting behavior of the separation of duties principle becomes less when there are more resources available. In the situation with only two resources the principle excludes 50% of the resources from handling the second task, where with five resources, see Fig. 3, only 20% of the resources is excluded from the handling the case.

The case handling resource dependency has an equal impact on the resource utilization as the separations of duties resource dependency in the scenario with only two resources, see Fig. 3, where both lines overlap but deviate from the line without resource dependencies. When increasing the number of resources the effect does not diminish as with the separation of duties resource dependency but increases significantly. In the scenario with five resources the utilization drops to 0.8 when it is expected to be 1.0. If the number of resources is further increased to eight resources then the utilization drops further to 0.7 instead of the expected 1.0. In contrast to the separations of duties resource dependency, the case handling resource dependency shows the same effect with only two resources as the separation of duties resource dependency, but in stead of a decrease in the restricting behavior the restricting behavior increases. For the scenario of two resources the principle excludes 50% of the resources from handling the second task. Where with five resources the restricting behavior has increased to excluding 80% of the resources to handle the second task. This translates to a heavily decrease of resource utilization in the model compared to the expected performance of the model. Therefore the waiting times in the model explode, since there is a lack of resources for processing these cases. Next to the lack of availability of resource because of the resource dependencies there is a random allocation strategy applied for the resources. With a process aware resource allocation algorithm this effect can be lessened, but the resource restrictions still apply, so the lack of resource availability still has an effect on the performance of the process.

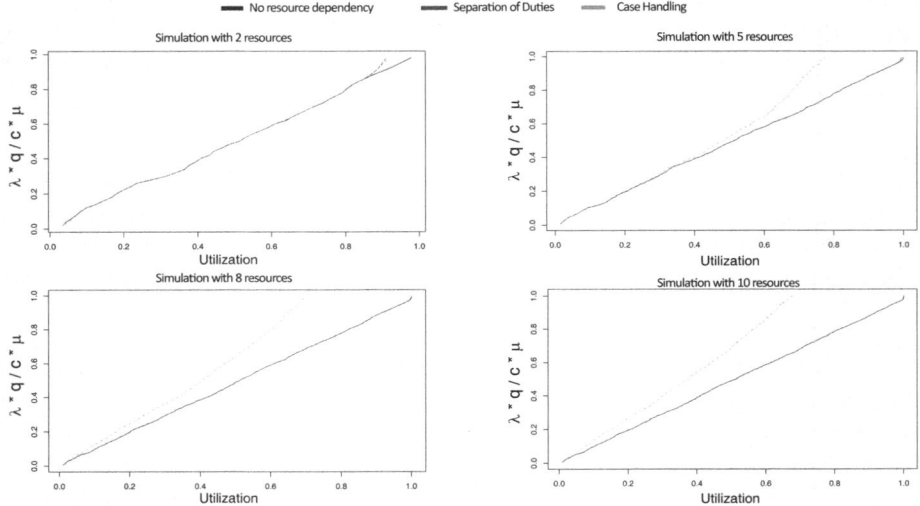

Fig. 3. Comparison with different number of resources

6   Conclusion and Future Work

This paper contributes to the field of business process management and business process simulation by providing a new simulation engine which provides the foundation for further incorporation of resource perspectives in business process simulation. In the related work section the lack of incorporation of certain aspects of business processes is shown. In the evaluation it is shown that the effects of not incorporating these aspects can lead to significant differences in performance metrics for the simulation. And since in practice companies aim for high utilization rates of 0.8 and above for their resources, these differences found are very relevant for the performance of the real world processes.

For future research the effect of the resource dependencies in larger processes can be studied to see how big the influence is on a large scale. Furthermore the use of advanced allocation algorithms for resources to cases can be investigated to obtain better performances of the model using the same set of resources. Also the resource assignment can be more formalized, as for example in Cabanillas et al. [2]. The effects of other extensions in the process model can be investigated further to show their impact on the performance of the process. For the discrete event simulation more mappings are possible, for example to map BPMN first to CPN Tools and then to discrete event simulation, here is chosen to map BPMN directly to discrete event simulation. Not all patterns from Russell et al. [18] and their interactions are described, which is a limitation of this paper. Also a limitation is that there is only looked at a single process instance, to increase the validity of the results more instances should be tested. Furthermore the simulation tool can be validated in future work by comparing it to other simulators, for example CPN Tools. In general more research should be conducted in the field of business process simulation, especially in combination with advanced resource perspectives. Thus this simulation engine is the first step in the direction of accurate simulation of complex real world processes.

Acknowledgment. The research described in this paper is supported by the Dutch Organization for Scientific Research as part of the DATAS project with grant number 438-15-507.

References

1. Andrews, T., et al.: Business process execution language for webservices (2003)
2. Cabanillas, C., Resinas, M., del-Río-Ortega, A., Ruiz-Cortés, A.: Specification and automated design-time analysis of the business process human resource perspective. Inf. Syst. **52**, 55–82 (2015)
3. Centeno, M.A.: An introduction to simulation modeling. In: Proceedings Winter Simulation Conference, pp. 15–22 (1996)
4. Dijkman, R.M., Dumas, M., Ouyang, C.: Semantics and analysis of business process models in BPMN. Inf. Softw. Technol. **50**(12), 1281–1294 (2008)
5. Dumas, M., La Rosa, M., Mendling, J., Reijers, H.A.: Fundamentals of Business Process Management, vol. 1. Springer, Heidelberg (2013)
6. Hlupic, V., Robinson, S.: Business process modeling and analysis using discrete-event simulation. In: Proceedings of the 30th WSC, pp. 1363–1370 (1998)
7. Jansen-Vullers, M.H., Netjes, M.: Business process simulation - a tool survey. In: Workshop on CPN Tools, Aarhus, Denmark, vol. 38, pp. 1–20 (2006)

8. Jensen, K., Kristensen, L.M., Wells, L.: Coloured Petri Nets and CPN Tools for modeling and validation of concurrent systems. Int. J. Softw. Tools Technol. Transfer **9**(3–4), 213 (2007)
9. Kelton, W.D., Sadowski, R.P., Sturrock, D.T.: Simulation with Arena. McGrawHill, New York (2004)
10. Kleinrock, L.: Queueing Systems, volume 2. Computer Applications, vol. 66. Wiley, New York (1976)
11. Kunze, M., Weske, M.: Signavio-Oryx academic initiative. In: BPM 2010 Demonstration Track 6 (2010)
12. Law, A.M., Kelton, W.D.: Simulation Modeling and Analysis, vol. 2. McGraw-Hill, New York (1991)
13. Mahajan, P.S., Ingalls, R.G.: Evaluation methods used to detect warm-up period in steady state simulation. In: Proceedings of the 36th WSC, Winter Simulation Conference, pp. 663–671 (2004)
14. OMG: Business Process Modeling Notation (BPMN) (2011). www.omg.org/spec/BPMN/2.0/PDF
15. Recker, J.: Opportunities and constraints: the current struggle with BPMN. Bus. Process Manag. J. **16**(1), 181–201 (2010)
16. Righter, R., Shanthikumar, J.G., Yamazaki, G.: Extremal service disciplines in single-stage queueing systems. J. Appl. Probab. **27**(2), 409–416 (1990)
17. Rozinat, A., et al.: Workflow simulation for operational decision support using design, historic and state information. In: Proceedings of BPM, pp. 196–211 (2008)
18. Russell, N., Ter Hofstede, A.H., Edmond, D., Van der Aalst, W.M.P.: Workflow resource patterns. In: BETA Working Paper Series WP 127, Eindhoven University of Technology, Eindhoven, The Netherlands (2004)
19. Stroppi, L.J.R., Chiotti, O., Villarreal, P.D.: A BPMN 2.0 extension to define the resource perspective of business process models. In: XIV Congreso Iberoamericano en Software Engineering (2011)
20. Ter Hofstede, A.H., Van der Aalst, W.M.P., Adams, M., Russell, N.: Modern Business Process Automation: YAWL and Its Support Environment. Springer, Heidelberg (2009). https://doi.org/10.1007/978-3-642-03121-2
21. Sarshar, K., Loos, P.: Comparing the control-flow of EPC and petri net from the end-user perspective. In: van der Aalst, W.M.P., Benatallah, B., Casati, F., Curbera, F. (eds.) International Conference on Business Process Management, pp. 434–439. Springer, Heidelberg (2005). https://doi.org/10.1007/11538394_36
22. Tumay, K.: Business process simulation. In: Proceedings of the 28th Conference on Winter Simulation, pp. 93–98 (1996)
23. Van der Aalst, W.M.P., Nakatumba, J., Rozinat, A., Russell, N.: Business process simulation. In: Brocke, J., Rosemann, M. (eds.) Handbook on Business Process Management 1, pp. 313–338. Springer, Heidelberg (2010). https://doi.org/10.1007/978-3-642-00416-2_15
24. Van der Aalst, W.M.P., Ter Hofstede, A.H.M., Kiepuszewski, B., Barros, A.P.: Workflow patterns. Distrib. Parallel Databases **14**(1), 5–51 (2003)
25. Van der Aalst, W.M.P., Weske, M., Grünbauer, D.: Case handling: a new paradigm for business process support. Data Knowl. Eng. **53**(2), 129–162 (2005)
26. Van Gorp, P.M.E., Dijkman, R.M.: A visual token-based formalization of BPMN 2.0 based on in-place transformations. Inf. Softw. Technol. **55**(2), 365–394 (2013)
27. De Vreede, G.J., Verbraeck, A., Van Eijck, D.T.: Integrating the conceptualization and simulation of business processes. Simulation **79**, 43–55 (2003)

28. Wohed, P., Van der Aalst, W.M.P., Dumas, M., Ter Hofstede, A.H.M., Russell, N.: Pattern-based analysis of BPMN (2005)
29. Zur Muehlen, M., Recker, J.: How much language is enough? In: Seminal Contributions to Information Systems Engineering, pp. 429–443. Springer, Heidelberg (2013)

Track III: Management

A Typology of Different Forms of Business Process Standardisation (BPS)

Kanika Goel$^{(\boxtimes)}$, Wasana Bandara, and Guy Gable

Queensland University of Technology, Brisbane, Australia
kanika.goel@hdr.qut.edu.au,
{w.bandara, g.gable}@qut.edu.au

Abstract. Organisations are increasingly adopting a process-centric view in order to compete and thrive. Business Process Standardisation (BPS) is a strategy for improved efficiency and effectiveness of business processes. BPS approaches are known to vary in practice, but are ill-understood. Through inductive content analysis of 18 published BPS case studies we identify three key decisions of the BPS approach: (i) Origin of standardisation (de facto or de jure), (ii) Optimisation of the master process (yes or no), and (iii) Choice of master process (internal exemplar, internal best-of-breed, or external exemplar) thus yielding twelve ($2 \times 2 \times 3$) alternative BPS types. This typology can serve as a useful tool for researchers investigating the BPS concept and may provide insight for practitioners when selecting an appropriate form of BPS. Further development, extension, and evaluation of the typology are suggested as future research.

Keywords: Typology · Business Process Standardisation · Case study
Inductive content analysis

1 Introduction

"Standardization is the activity of establishing and recording a limited set of solutions to actual or potential matching problems directed at benefits for the party or parties involved balancing their needs and intending and expecting that these solutions will be repeatedly or continuously used during a certain period by a substantial number of parties for whom they are mean" [1]. Business Process Standardisation (BPS) has been recognised as a key mechanism for achieving operational process optimisation, and is becoming integral to Business Process Management practice (BPM) [2]. BPS has been shown to positively impact business performance [3, 4], as evidenced in time, cost, and quality matrices [5]; it can also facilitate streamlining, automating, or outsourcing of business processes [6]. Globally, companies both private (e.g. [7]) and public (e.g. [8]) are making substantial investments in standardizing their business processes.

BPS can arise in different forms, which results from the different choices made in BPS design and execution. Understanding these underlying decision options and the different forms they result in, will assist in bringing more clarity to the concept of BPS and enable better BPS design, implementation, and decision-making. However, this aspect of business process standardisation has received limited attention. This study

© Springer Nature Switzerland AG 2018
M. Weske et al. (Eds.): BPM Forum 2018, LNBIP 329, pp. 179–193, 2018.
https://doi.org/10.1007/978-3-319-98651-7_11

asks 'What are the different forms of Business Process Standardisation?' and uses the literature to identify the prominent differentiating decisions in order to develop a typology for Business Process Standardisation. Typologies provide a useful framework to explain outcomes [1]; they aid analysis and provide a means for comparing and contrasting classes of a phenomenon (Gregor [2]). A typology of BPS can help unveil different forms of standardisation to determine which form is suitable in different contexts. First, we briefly discuss the concept of BPS; next, we present the research approach; this is followed by the findings, and then a related discussion.

2   Introductory Overview of Business Process Standardisation (BPS)

For business processes to be standardised, there must be a 'standard' to adopt. According to ISO [3]:

"Standards are documents, established by consensus and approved by a recognized body that provides, for common and repeated use, rules, guidelines or characteristics for activities or their results, aimed at the achievement of the optimum degree of order in a given context".

When establishing a process standard, a 'Master process' (a process that becomes the reference point) must be determined [4], and there are several ways in which to do this. Firstly, a master process may be derived internally by combining the better practices of the process variants. A process variant is "an observed or documented business process with a specific variation of at least one of the elements (inputs, outputs, enablers, guides and sequence of activities) for a defined part of the overall process" [5]. Another way a Master process can be developed is by adopting an internal end-to-end process in its entirety [6]. Finally, a Master process may be a reference point external to the organisation [7]. This concept of the 'Master process' is referred to variously as an 'Archetype process' (Muenstermann and Weitzel [4]) or a 'prototype standard' (Muenstermann, Eckhardt [8], Stetten, Muenstermann [9]); all referring to a reference point that is derived internally or externally, against which all other process variants are to be standardised.

Muenstermann and Weitzel [4] suggest particular characteristics of a Master process that are essential for it to become a standard (see Table 4, p. 10 of their paper). They explain that the chosen master process must be 'well documented' and 'modularized' (which is the process of subdividing a process into meaningful sub-processes and steps), with 'specificities' clearly 'isolated'. Specificities refer to those steps that cannot be undertaken in the same way in all process instances. Such steps need to be sequestered and minimized in a good process standard [4]. At the time of its adoption, a Master process might also be improved (beyond selecting the best internally or externally existing end-to-end master process, or combining existing internal best practice modules), to reflect known best practices and thereby ensuring the standard is 'the best known execution process' [4, 10, 11]. Once this process has been devised, it is endorsed and becomes standardised, by unifying the variants of the process in line with the standard.

3   Research Approach

This research adopted a multi-phased approach to build a typology of BPS. A detailed coding rule book (following [12]) was prepared, outlining how the data was to be captured, stored, updated, and analysed; this was strictly adhered to. Data was analysed primarily by one coder, and then followed by detailed coder-corroboration sessions with a second coder. In the case of any discrepancies or disagreements, the original data was revisited to jointly resolve the issue. Figure 1 demonstrates the approach followed.

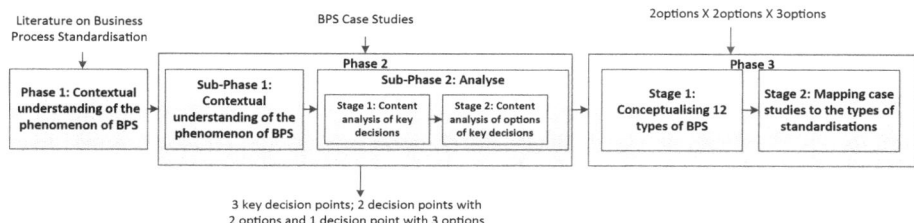

Fig. 1. Phases and stages involved in research approach

In Phase 1, we reviewed existing BPS literature to understand the key stages involved in the process of standardization (outcome presented in Sect. 2). Such provided a sound contextual understanding of what BPS is, including some early elicitations of choices organisations have (or decisions that organisations make) during design and implementation of BPS. Phase 2, consisted of two sub-phases. First, we sought for published BPS case studies to form this study's empirical base. Given that documented case studies provide vivid descriptive information, and with the goal of been able to use such case details of BPS as an empirical base for this study, Phase 2 embarked on a structured search and analysis of published BPS case studies, following the guidelines of Bandara et al. [13].

As current literature around BPS is scarce and disparate [14], we aimed to retrieved all BPS articles and then review them to extract case studies. A broad range of search strings[1] were employed, to locate full text, peer-reviewed journal and conference articles in English. Following search strategies of similar literature reviews [15], the databases JSTOR, IEEE, Emerald, ABI/Inform, Science Direct, ProQuest, and Gartner were used. The retrieved papers were quality and relevance checked, resulting in 38 papers about BPS, which increased to 44 papers with forward and backward searching. These papers were reviewed in detail to identify those that had case details providing a descriptive overview of the implementation of BPS. This resulted in 18 case studies and 22 cases (as some papers had more than one case study). These cases were then subject to content analysis to elicit the different decisions made (as explained directly or

---

[1] Some examples include: Process and standard* in abstract, title and keywords; "Business Process" AND standard* in abstract, title and keywords; Process AND standardization in in abstract, title and keywords; "Business Process" AND standardization in in abstract, title and keywords.

implied in the case studies) that characterized the diverse BPS efforts described in the cases. The analysis was inductive and involved two stages (Stage 1 and 2). In Stage 1 the aim was to distil key decisions undertaken by the organisations during their standardization initiative. In Stage 2, the content was revisited to synthesise various options for each of the decisions extracted in Stage 1[2]. The aim was to find distinct decisions that characterised the different BPS implementations, and to then delineate the patterns in which these decisions manifested. The outcome of this Phase was three decisions (see Sect. 4.1) and associated options (Fig. 1).

Phase 3 also had two stages. In Stage 1 we aimed to conceptualize the different forms of BPS by generating categories that simply combined the decisions options from Phase 2; which resulted in 2 * 2 * 3 = 12 potential forms of BPS. Next, we revisited the case studies to map them against each of the potential types. This resulted in instantiation of 8 of the conceptualised types of BPS. Though not supported by empirical evidence (i.e. case instantiations) the other 4 types are also explained and presented here using a hypothetical case for illustrative purposes. This was done, given these BPS types' logical relevance and also potential limitations of the current pool of cases selected for this analysis (i.e. only based on published BPS cases in academic outlets).

4   Study Findings

This section first introduces the resulting decisions that contribute to differentiating BPS implementations, explaining the potential options within each decision. Following this, different types of BPS are discussed (which are derived by the different combinations of these decisions options).

4.1   Key Decisions that Differentiate the Different Forms of Business Process Standardisations

Three key decisions that potentially explained the different forms of Business Process Standardisation were uncovered.

Decision 1 (D1) Origin of Standardisation: This decision relates to the nature of standardisation, which results from the different triggers causing the organisation to undergo BPS. Kwon [16] discusses two possible options for the origin of standardisation: de facto and de jure. He explains: "De Facto standardisation emerges spontaneously and informally, whereas de jure standardisation is usually issued from administrative procedures and enforced by authorities that have some regulatory power" [16]. Therefore, the de jure origin of standardisation is formalised and specifies

---

[2] For instance, incorporation of best practices is a decision to be taken by the organisation during standardization, and was distilled from literature as an outcome of Stage 1; whereas, the options of the decision (yes or no), were extracted and synthesised in Stage 2.

how things need to be done. A formal project is mandated and then initiated by upper management. In contrast, de facto standardisation emerges organically, led by employees as a means to try to improve their day-to-day practice.

Decision 2 (D2) Optimisation of Master Process: This decision relates to the inclusion of best practices in the Master process, when formulating the standard in the process of standardisation (refer to Sect. 2 for an overview of BPS). The literature conveys how the integration of best practices depends on organisational goals of standardisation: only when an organisation wishes to have the processes standardised against an improved version of the Master process are best practices are integrated; otherwise, the organisation may choose not to.

Decision 3 (D3) Type of Master Process: When standardising a process, a Master process is sought first, before the process is standardised (see Sect. 2 for further details). This dimension relates to the type of Master process chosen. The literature clearly outlines two possible forms of internally-derived Master processes and one external type. The first internal type uses a reference process that is chosen from within the organisation [17, 18]. This internal Master process can then be a complete end-to-end single process chosen from within the organisation [6] or an amalgamation of modules (best-of-breed) of internal variants of the process [4]. The analysis (see below) confirmed that there may be an external Master process, against which the organisation wishes to standardise its processes. For example, given the large volume of best process practices that has been captured through decades of research forming reference process models in different domains [19] and more recently been available through structured process model repositories [e.g. 20], organizations can refer to these when considering external input to form the Master process for their BPS efforts.

4.2   Different Forms of Business Process Standardisation

Of the three decisions introduced above, D1 and D2 have two options each, and D3 has three options. And when looking at the combinations this can result in; $2 \times 2 \times 3$—it points to twelve potential types of Business Process Standardisation, as presented in Table 1. Of these twelve types, eight were instantiated by the literature and four (Type 6, 8, 11 and 12) were not. Column 1 of Table 1 populates the different forms of BPS (referred to as BPS 'Types') in accordance with the decisions and the options for them uncovered in the literature. Columns 2, 3, and 4 list the three decisions and the options within, as evidenced in literature (as explained in Sect. 4.1). Column 5 lists the total number of cases in which the type of standardisation was evidenced. Column 6 lists the references where the types of standardisation were uncovered. When the reference had more than one case study, then the reference has been suffixed with a case number (i.e. Case 1, Case 2, and so on) to reflect the corresponding cases described in the same paper. These results were carefully checked and reviewed by two coders for quality assurance.

Table 1. Types of Business Process Standardisation (BPS)

	D1	D2	D3				
			Type of Master Process				**Cases**
Type	Origin	Optimise	Internal Exemplar	Internal best-of-breed	External exemplar	No of Cases	
1	De jure	Yes	X			8	(Agnar, Harry, & Mariano, 2004; Kwon, 2008; Manrodt & Vitasek, 2004; Muenstermann et al., 2009; Müenstermann et al., 2010; Rahimi et al., 2016, Case 3; Rosenkranz et al., 2010, Case 2; Schafermeyer et al., 2010, Case 1)
2				X		2	(Muenstermann, Moederer, & Weitzel, 2010; Müenstermann & Weitzel, 2008)
3					X	1	(Wessel, Ribbers, & Vries, 2006)
4		No	X			3	(Afflerbach et al., 2016; Rosenkranz et al., 2010, Case 3; Stetten et al., 2008)
5				X		3	(Kettenbohrer, Beimborn, & Kloppenburg, 2013a, 2013b; Rahimi, Møller, & Hvam, 2016, Case 2)
6					X	NA	
7	De facto	Yes	X			1	(Roubert, Beuzelin-ollivier, Hofmann-amtenbrink, Hofmann, & Hool, 2016)
8				X		NA	
9					X	1	(Kauffman & Tsai, 2010)
10		No	X			2	(Rahimi et al., 2016, Case 1; Schafermeyer et al., 2010, Case 2)
11				X		NA	
12					X	NA	

Type 1: De jure standardisation, with an internal exemplar Master process and optimisation of Master process

This type of standardisation is suitable to organisations where standardisation is a result of formal authority and procedures to achieve consistency in processes, and where standardisation efforts are related to continuous improvement. Organisations working towards such standardisation would have multiple variants of the process to be standardised and are committed to process improvement. This is why an internal exemplar Master process is sought and then optimised with best practices for the BPS initiative. For example, in the case study outlined by Muenstermann, Eckhardt [8] the organisation 'VISION' launched a project to standardise its recruitment process across all autonomous divisions working under the umbrella of the company. VISION pre-selected the recruiting process of the headquarters as the Master process and then enhanced it, using the insights gained from three large competing organisations regarding strengths and weaknesses of the process. This helped VISION to standardise their processes against an industry-identified best practice process, assisting the organisation in achieving the specific goals it was committed to. Eight cases were instantiated as Type 1; in other words, this is quite a popular form of standardisation employed by organisations. Our analysis suggests that this type of standardisation is suitable for organisations with mature processes (from which a Master process can be chosen), and a goal of standardising against an improved version of the Master process.

Type 2: De Jure standardisation, with a best-of-breed internal Master process and optimisation of Master process

This type of standardisation is suitable for organisations where the BPS initiative is a result of a lack of consistency in the processes; these organisations are aiming for continuous improvement. Organisations undertaking this form of BPS document all the variants as a part of the formal procedures and merge best practice modules of variants of the same process to obtain a Master process.

The best-of-breed internal process is further improved by integration of industry best practices, enabling the processes to be standardised against a best practice process that was derived by amalgamating a range of internal best practices and the enhancing it with external best practice. In the case study outlined by Muenstermann and Weitzel [4], a multinational firm 'Dream' (real name kept anonymous by authors) launched a program for BPS with a desire to reduce costs and work on continuous improvement. The organisation had a number of process variants and used Type 2 standardisation to achieve their goals. This type of standardisation hence works best in organisations with mature processes, and with multiple variants. The objective to implement standardisation is consistency and process improvement. Two cases instantiated this form of standardisation.

Type 3: De Jure standardisation, with an exemplar external Master process and optimisation of Master process

This type of standardisation is suitable for organisations where standardisation is formally introduced due to observed inconsistency in processes, with a desire to standardise against an external best practice process, and a goal of continuous improvement. Type 3 particularly relates to organisations that do not have competent processes or are in industries in which there is a need to abide by external standards (such as IT, law etc.). In their paper, Wessel, Ribbers [21] discuss a standardisation initiative across a financial firm with over 50 divisions around the world. The firm had integrated an information system, 'PeopleSoft', to standardise their business processes. This helped the firm to ensure that the data privacy and protection regulations were followed in the same way across all the divisions. Further, there was a focus on continuous improvement to ensure that the Human Resource policies are best practices, enabling the organisation to perform better. This type was instantiated by one case study.

Type 4: De Jure standardisation, with an internal exemplar Master process and no optimisation of Master process

This type of standardisation is suitable for organisations that want to standardise their processes, where the process standardisation efforts are not related to any continuous improvement, and the aim is mere consistency of processes based on existing/current internal practices. This form of BPS is common with the rise of globalisation as the desire and need for greater consistency in services/operations grows. These organisations require a dedicated project to standardise the processes; this involves committing certain resources to achieve the target outcome of overall consistency. Formal official procedures need to be followed and final approval for the standard needs to be obtained. A responsible person needs to be appointed for the BPS initiative [22]. Once a formal procedure for standardisation has been initiated, a decision to choose the Master process needs to be made: internal or external. Organisations with mature internal processes may choose to have an entire internal (e.g. a headquarters' process) as the Master process, in this type. So this one internal process becomes the point of reference, which is then used to standardise the other processes.

This type of standardisation was evident in the Stetten, Muenstermann [9] case study of 'Future' (name of the case has been anonymised in the paper). 'Future' has three autonomous divisions, each responsible for their own results. In the past, different information systems were being used for the same goal: recruitment of candidates. 'Future' wanted the recruitment process to become consistent across all divisions, so it launched a standardisation project (De Jure) in 2004. The aim was to have maximum internal and external transparency, with no commitment to process improvement. Since the processes at 'Future' were mature, the headquarters process was chosen as the Master against which to measure for the standardisation process. This type of standardisation has been instantiated by three case studies from the pool of cases analysed.

Type 5: De Jure standardisation, with a best-of-breed internal Master process and no optimisation of Master Process

This type of standardisation is applicable to organisations that have a relatively greater number of process variants and hence want to standardise the process, where the desire is consistency of processes across the organisation, without any goal for continuous improvement. When compared to Type 4, the difference here is that the organisation is likely to have a number of process variants, and modules of such variants are merged to obtain the Master process to be used for standardisation. This type of standardisation was reflected in the case of Lufthansa Technik (LHT) in Germany [23] where the goal was to standardise the process that provides quality assessment of suppliers and supplier-related products. There were several working variants of the same process, with modules that were considered best practice internally by the organisation. Such pockets/modules were extracted and then amalgamated to retrieve the Master process, which was then used for standardisation. This type of standardisation has been instantiated by three case studies from the pool of cases analysed.

Type 6: De Jure standardisation, with an external Master process and no optimisation of Master process

This type of standardisation was not instantiated by any case study. However, we believe this type of standardisation can exist. An organisation may formally launch a standardisation initiative and choose to use an external standard as the Master process, without any need to optimise it. For example, University A may wish to standardise its student enrolment process. They may feel that their own current processes (or any variants) are not competent, but on the other hand, have confidence in the student enrolment process of University B because of immense positive feedback (or other evidence). In such a case (assuming that University A has access to University B's, process details, either in the public domain or through a structured review), University A may launch a dedicated standardisation project to standardise its student enrolment process, using University B's method as the Master process. Since University A is satisfied with the execution of the student enrolment process of University B, it may decide not to enhance it with integration of best practices. Hence, this type is suitable for organisations that desire consistency of operations, but do not consider their own internal processes sufficient to be used for a standardisation initiative, but may want to standardise their processes against a best-known external process.

Type 7: De Facto standardisation, with an exemplar internal Master process and optimisation of Master process

This type of standardisation is suitable for organisations where standard processes are required for efficiency and people are encouraged to keep their practices as uniform as possible across the organisation. Further, this type of firm would have competent mature processes, leading them to use an internal Master process and commit to improvements that will lead to the optimisation of a Master process. Standardisation in such firms assists people to do their tasks well, which is why observed practices start to become a standard. This type of standardisation was somewhat evident in the case study by Roubert, Beuzelin-ollivier [24], where a firm working in nanotechnology realised its need for a standard process related to nanoparticle experimentation. The current set of guidelines the company used to conduct experiments became the Master process. This was further enhanced by collating information from external protocols followed for tasks related to nanoparticles. Mutual consensus was gained on the integration of such best practices. This assisted in developing standard practices with nanoparticles across the firm. One case instantiated this type of standardisation.

Type 8: De Facto standardisation, with a best-of-breed internal Master process and optimisation of Master process

This type of standardisation was not instantiated by case studies; however, we believe it could happen. When the employees in an organisation experience confusion and they are over-worked, they may initiate the need for consistency. They may decide to standardise the processes (de facto) using the observed practices. Since many variants of the process may exist, causing inconsistency, the organisation may use best practices from variants to form an internal Master process and then refine it further with other industry best practices, to achieve the best possible outcome. For example, University A may have many variants of the student recruitment process across faculties, causing inconsistency in data stored. As a result, the employees may attempt to standardise the process by selecting different parts (or modules) of the recruitment process from variants to form a new Master process. The Master process may be further optimised by integrating practices from recommended best practice student recruitment processes by the national education board. However, this form may not be widely evident: for it to be effective and successful, the employees need to have a deep desire for continuous improvement and an understanding of process management.

Type 9: De Facto standardisation, with an exemplar external Master process and optimisation of Master process

This type of standardisation is suitable for organisations that cannot function efficiently without having standard practices, but in which no standardisation is mandated. In these cases, the staff themselves initiate the standardisation process, as they see it as helping them to do their job better (which is also why De facto standardisation happens). This form exists in industries where there is fierce competition, creating a need to maintain standard practices across the entire industry—examples include IT, law, and medicine.

This type of business process standardisation was evident in the case study outlined by Kauffman and Tsai [7] in which an IT firm adopted de facto standardisation to ensure the technology sold to the consumers is of consistent quality. The employees considered standardisation necessary: without it, the customer may not be happy. For example, software vendors need to have standardised practices to launch enterprise software solutions, so the employees started looking for ways to do this. An industry-wide-practiced standard was chosen, which was subject to continuous refinement. Adopting such a standard assisted the IT firm to maintain consistent practices across the firm and also compare their performance across the industry. Only one case was instantiated for this type of standardisation.

Type 10: De Facto standardisation, with an exemplar internal Master process and no optimisation of Master process

This type is suitable for organisations that have mature (processes with intricate details), yet inconsistent processes, but desire more consistency to fulfil the tasks with reduced time, energy, and costs. This type was evident in case study 1 described by Schafermeyer, Grgecic [25] about the German telecommunication provider (TCSP) that started using a software as a part of its client order process, and then went on to adopt this as a standard. Following this, an in-house software was developed to achieve the desired standardisation of the client order process. Since the aim was solely to have a consistent client order process across all the locations of the organisation, the Master process (the internally designed Master process that was embedded in the system) was not optimised. One case was instantiated for this type of standardisation.

Type 11: De Facto standardisation, with a best-of breed internal Master process and no optimisation of Master process

This type of standardisation was not instantiated by any case study, though we believe it can exist. It is suitable for organisations where consistency in processes is crucial and there are multiple variants of the same process in the organisation. There is no goal of continuous improvement. For example, University A may have variants of student recruitment process across faculties, resulting in inconsistencies in data stored, which then, in turn, may cause bad decision making. Therefore, to make things easier for the staff, the staff members may decide to work together to understand the different variants; once variants have been determined, the best practices may be combined to form the Master process. The recruitment process may then be standardised against that Master process. Since the goal is consistency and (not to have an optimal recruitment process), the Master process is not optimised further.

Type 12: De Facto standardisation, with an exemplar external Master process and no optimisation of Master process

This final type of standardisation was not instantiated by the case studies, either, but, again, we believe it may happen. For example, the employees of University A may be tired of managing the different ways they process the payroll. Some employees may ask for signatures of the concerned, for instance, while others may not, resulting in confusion and inconsistency in the data fed into the systems. As a result, the employees may internally discuss and initiate standardisation of the payroll process. In order to do this, they may adopt the guidelines of the payroll process of some other university, which they believe to have better operations than the one in their own organisation; using that information, they work to standardise the process at University A.

5   Discussion

The aim of this paper was to review the existing literature around Business Process Standardisation (BPS) in an attempt to build a preliminary BPS typology. Analysis of content revealed three decisions (D1, D2, and D3). An organisation seeking the benefits of BPS needs to first decide whether to lead standardisation (de jure). Or, standardisation may also occur organically, as different parts of the organisation become aware what is working well and what not (de facto). Next, there are three options for deciding the Master process (internal exemplar, best-of-breed internal, external exemplar). Finally, an organisation may decide optimising the Master process by taking into account best (industry) practices. A total of twelve types of BPS were conceptualised, considering the options for the three decisions. Eight types were instantiated by the case studies. However, we do believe that the other types could also exist and have provided hypothetical scenarios (based on University A) to illustrate these alternative types. Of all types, Type 1 (De Jure standardisation, with an exemplar internal Master process and integration of best practices) was found to be the most popular. BPS initiatives involve considerable time, money, and resources, which is why companies may launch a formal project to standardise a process. Moreover, a best-known internal process may be chosen (as it comes with advantages, such as having access to the process details and the fact that the process parts been formed within the same context), and then may be refined with best (industry) practices. Considering these aspects, it is not surprising to find Type 1 to be the most common.

With growing significance of BPM and increasing globalisation, the concept of BPS is also gaining prominence [14]. Greater numbers of organisations are adopting

BPS as a mechanism to achieve consistency in operations and obtain the benefits of reduced time and costs, while improving service quality. While traditionally the objective was to standardise against the best practice process, with a goal of process improvement, today there are organisations that desire mere consistency of processes, driven by the need to provide consistent customer service. The objective is to have uniform, rather than best practice processes functioning within the organisation. This is why in some cases the Master process is not subject to improvement through integration of best practices (Type 1, 2, 3, 7, 8, and 9). Another reason for not optimising the Master process can be the organisation having competent and mature processes, which is why further improvement is not sought. Further, a specific type of Master process, best-of-breed internal, has emerged. This is particularly true for organisations that are large and have multiple variants of the same process. In such cases, the organisation may consider having sections of the variants amalgamated together to obtain the best practice internal process, against which the variants are then standardised (Type 2, 5, 8 and 11). Further, organisations that are new, with incompetent processes, may also choose to have standard processes. In such cases, an external Master process is found suitable to be adopted (Type 3, 6, 9 and 12). Adopting an external Master process is also evidenced as applicable to organisations where there is a need to maintain consistency with industry-wide practices. For example, a banking organisation has to follow certain rules and regulations practiced across the industry in order to sustain its place in the market. Furthermore, on understanding the options of D2, we believe another option is possible which is combining an external standard with internal practices to form a best-of-breed Master process, which is not entirely internal. For instance, a university while standardising their recruiting process may use activities of an external best practice process and amalgamate it with the activities of the internal process to arrive at the best-of-breed Master process. However, the literature does not instantiate such option.

Further, the origin of standardisation may differ for organisations. Standardisation may be a result of a formal project initialised by the authorities in an organisations resulting in a dedicated team or resources (de jure), or may not be formal in nature (de facto) and emerge organically; led by employees as a means to try to improve their day-to-day practices. De jure standardisation seems to be more common (nine cases instantiated this), which is understandable, as considerable resources are spent in such initiatives. However, employees have started recognising the need for having standardised processes that can aid them in reduced rework, time, cost, resources, and efforts, and are more and more often encouraged to engage in continuous improvement through standardised practices. This is why de facto standardisation has started gaining significance, especially in industries where it is difficult to function without standard practices, such as the medical industry.

Three decisions revealed in the literature assisted us in understanding the different types of standardisation that may take place. Options within a decision can influence the state of other decisions. For instance, de jure BPS is more likely to proceed with optimisation (if not immediately, at least as a follow-up phase) because optimisation efforts are more likely to be supported/resourced by management in de jure than in de facto contexts. This also brings forth the plausibility of shifting from one type of standardisation to another, as BPS itself is a continuous effort, which is reviewed and repeated on a periodic basis (i.e. companies may update their standard process and then unify the variants with the standard process). In such instances, the organisation may have initially conducted one type of BPS, but in the next iteration they may move to the next. For example, the success of de facto BPS can lead to de jure BPS. Or a company may deploy Type 4 (De Jure standardisation, with exemplar internal Master process and no optimisation of Master process) first, but in the next iteration may choose to switch to Type 1 (De Jure standardisation, with exemplar internal Master process and optimisation of Master process) to further enhance the Master process against which the variants are eventually standardised. An understanding of these different types of standardisation, as revealed in this paper, is significant: it brings further clarity to the concept of BPS. These different types assisted in building a preliminary typology for BPS, as discussed in this paper, and provided a framework to explain different outcomes [1]. According to Gregor [2], typologies help in understanding similar yet different forms of a phenomenon, in this case BPS, enabling better understanding. This understanding can serve as a useful tool for researchers investigating the BPS concept and also provides insights for practitioners to select the appropriate form of BPS, for their specific purposes and circumstances.

We acknowledge the limitations inherent in the research approach and analysis here. Our search outcomes could have potential limitations, as each and every article related to BPS may not have been retrieved. It is recognized that 22 cases were used to build this typology, and there may be other published BPM cases that were not retrieved. Further, even though two coders were involved in the coding process, it is possible that some relevant information may not have been coded. However, this was minimised by conducting multiple rounds of coding and having continuous corroboration sessions. This qualitative research was built on content analysis assisting in forming themes, which may at times appear to be subjective and lacking transparency in how the relevant themes were developed, but we have tried to address this by providing an overview of the data analysis. We also do not claim that the BPS decisions presented here that formed the dimensions characterizing BPS are the only ones - they were, however, the ones derived from the extracted cases. We suggest future researchers could validate and further build upon this preliminary typology of BPS. Given the BPS choices depend on context, researchers may conduct BPS case studies to further understand the organisational requirements and contexts related to each BPS type. Diverse organisational impacts based on the type of BPS need to be explored. Conducting further empirical work to validate the typology, research on the different success factors and impacts from these different BPS types, and how they may differ in their designs and implementations would be beneficial to progress the outcomes presented in this paper.

6   Conclusion

In this paper we used 22 BPS cases to understand the different forms of Business Process Standardisation (BPS). Content analysis revealed three decisions: (D1) origin of standardisation, (D2) Optimisation of Master process, and (D3) Type of Master process. D1 and D2 were evidenced to have two options and D3 had three options; the combination, resulting in twelve different types of BPS. Of these twelve types, eight were instantiated by the cases and the others were explained using hypothetical cases. The three decisions assisted in developing a preliminary typology for BPS. The typology presented herein provides a good understanding of the concept of BPS, and can provide a solid foundation for future researchers to unravel other potential types and further insights of BPS. The typology also has practical significance as it enables organisations to select the right type, given their organisational requirements. Future researchers are encouraged to build this typology and accumulate knowledge in a systematic manner, enlightening both professional and academic practice.

References

1. Doty, D.H., Glick, W.H.: Typologies as a unique form of theory building: toward improved understanding and modeling. Academy of Management. Acad. Manage. Rev. **19**(2), 230 (1994)
2. Gregor, S.: The nature of theory in information systems. MIS Q., 611–642 (2006)
3. ISO, Standardization and related activities — General vocabulary, I.O.f. Standardization), Editor (1996)
4. Muenstermann, B., Weitzel, T.: What is process standardization? In: International Conference on Information Resources Management (CONFIRM) (2008)
5. Zellner, P., Laumann, M., Appelfeller, W.: Towards managing business process variants within organizations – an action research study. In: 2015 48th Hawaii International Conference on System Sciences (HICSS) (2015)
6. Muenstermann, B., Moederer, P., Weitzel, T.: Setting up and managing business process standardization: Insights from a case study with a multinational e-commerce firm. In: 43rd Hawaii International Conference on System Sciences (2010)
7. Kauffman, R.J., Tsai, J.Y.: With or without you: the countervailing forces and effects of process standardization. Electron. Commer. Res. Appl. **9**(4), 305–322 (2010)
8. Muenstermann, B., Eckhardt, A., Weitzel, T.: The performance impact of business process standardization. Bus. Process Manage. J. **16**(1), 29–56 (2010)
9. Stetten, A.V., et al.: Towards an understanding of the business value of business process standardization - a case study approach. In: Americas Conference on Information Systems (2008)
10. de Vries, H.J.: Best practice in company standardization. Int. J. IT Stand. Stand. Res. **4**(1), 62–85 (2006)
11. Ungan, M.C.: Standardization through process documentation. Bus. Process Manage. J. **12**(2), 135–148 (2006)
12. Saldana, J.: The Coding Manual for Qualitative Researchers. SAGE (2012)
13. Bandara, W., et al.: Achieving rigor in literature reviews: insights from qualitative data analysis and tool-support. Commun. Assoc. Inf. Syst. **34**(8), 154–204 (2015)
14. Muenstermann, B.: State of the art of BPS research. In: Business Process Standardization: A Multi-Methodological Analysis of Drivers and Consequences, pp. 29–118. IGI Global, Hershey (2015)
15. Romero, H.L., et al.: Factors that determine the extent of business process standardization and the subsequent effect on business performance. Bus. Inf. Syst. Eng. **57**(4), 261–270 (2015)

16. Kwon, S.-W.: Does the standardization process matter? A study of cost effectiveness in hospital drug formularies. Manage. Sci. **54**(6), 1065–1079 (2008)
17. Afflerbach, P., Bolsinger, M., Röglinger, M.: An economic decision model for determining the appropriate level of business process standardization. Bus. Res. **9**(2), 335–375 (2016)
18. Muenstermann, B., Eckhardt, A., Weitzel, T.: Join the standard forces - examining the combined impact of process and data standards on business process performance. In: 42nd Hawaii International Conference on System Sciences, HICSS 2009 (2009)
19. Fettke, P., Loos, P.: Classification of reference models: a methodology and its application. Inf. Syst. e-Business Manage. **1**(1), 35–53 (2003)
20. La Rosa, M., et al.: APROMORE: an advanced process model repository. Expert Syst. Appl. **38**(6), 7029–7040 (2011)
21. van Wessel, R., Ribbers, P., de Vries, H.: Effects of IS standardization on business process performance: a case in HR IS company standardization. In: Proceedings of the 39th Annual Hawaii International Conference on System Sciences (HICSS 2006) (2006)
22. Rosenkranz, C., Seidel, S., Mendling, J., Schaefermeyer, M., Recker, J.: Towards a framework for business process standardization. In: Rinderle-Ma, S., Sadiq, S., Leymann, F. (eds.) BPM 2009. LNBIP, vol. 43, pp. 53–63. Springer, Heidelberg (2010). https://doi.org/10.1007/978-3-642-12186-9_6
23. Kettenbohrer, J., Beimborn, D., Kloppenburg, M.: Developing a procedure model for business process standardization. In: International Conference on Information Systems, Milan (2013)
24. Roubert, F., et al.: "Nanostandardization" in action: implementing standardization processes in a multidisciplinary nanoparticle-based research and development project. Nanoethics **10**(1), 41–62 (2016)
25. Schafermeyer, M., Grgecic, D., Rosenkranz, C.: Factors influencing business process standardization: a multiple case study. In: 2010 43rd Hawaii International Conference on System Sciences (HICSS) (2010)

How to Put Organizational Ambidexterity into Practice – Towards a Maturity Model

Maximilian Röglinger, Lisa Schwindenhammer,
and Katharina Stelzl[(✉)]

FIM Research Center, University of Bayreuth, Bayreuth, Germany
{maximilian.roeglinger, lisa.schwindenhammer,
katharina.stelzl}@fim-rc.de

Abstract. Organizational ambidexterity (OA) is a vital capability for surviving in dynamic business environments by simultaneously pursuing exploitation, i.e., continuous streamlining of business processes, and exploration, i.e., radical innovation of products, services, and processes. During the last years, OA knowledge has continuously matured, comprising insights into performance outcomes, antecedents, and moderators. However, there is a lack of guidance on how to put OA into practice. Addressing this challenge, our research is geared toward the development of an organizational ambidexterity maturity model (OAMM) using a design science research approach. Our OAMM follows a prescriptive purpose of use, helping organizations select actionable practices. To develop our maturity model, we first reviewed the general OA literature to identify actionable practices. Second, we built on the six core elements of BPM, i.e., strategic alignment, governance, methods, information technology, people, and culture, to structure identified practices. Third, we used card sorting to assign practices to maturity levels. We evaluated our OAMM with respect to general design principles for maturity models. Our work lays the foundations for the structured development of OA capabilities and for future research in this area.

Keywords: Organizational ambidexterity · Exploitation · Exploration
Maturity model · Capability development · BPM capabilities · Card sorting

1 Introduction

Organizational ambidexterity (OA) emerged as an essential capability to explain how organizations sustain success in dynamic and turbulent environments [30]. The enduring challenge lies in reconciling tensions between exploitation and exploration as two inseparable modes of organizational learning and change [26]. Yet a considerable number of organizations struggle in aligning and configuring the entire organization to solve these tensions and achieve a balance between exploitation and exploration [17].

To date, scholars have researched the outcomes, antecedents, and moderators of OA as three major streams in conceptual and empirical studies on ambidexterity [32, 38]. *Outcomes* relate to the positive performance effects that OA entails, e.g., in terms of sales growth, profitability, and operational performance [13, 19, 25]. *Antecedents* describe the elements or mechanisms of organizational design employed to achieve

© Springer Nature Switzerland AG 2018
M. Weske et al. (Eds.): BPM Forum 2018, LNBIP 329, pp. 194–210, 2018.
https://doi.org/10.1007/978-3-319-98651-7_12

balance between exploitation and exploration [44]. *Moderators* are all factors which influence the OA-performance linkage, e.g., competitive dynamics or firm size and age [38]. The benefits of OA as a competitive differentiator and precursor of long-term survival have been broadly recognized [13, 19, 25]. Therefore, research on OA antecedents investigates sequential and simultaneous approaches and thus, structural, contextual, and leadership-based antecedents [13, 32, 38]. Although their combination is considered beneficial [1, 38], the interrelations between different types of OA antecedents remain under-researched [38]. Hence, answers to the question of *how* to put ambidexterity into practice remain open and a lack of practical guidance persists [1, 31]. To address this gap, we seek to enhance prescriptive knowledge on OA capability development answering the following research question: *How to put OA into practice by systematically developing OA capabilities?*

In answering our research question, we adopt the design science research (DSR) paradigm [16] and develop an organizational ambidexterity maturity model (OAMM) as resulting artifact. Our OAMM serves a prescriptive purpose of use, assisting organizations in the development of OA capabilities based on actionable practices. They describe clear actions helpful to implement OA and thus assist the configuration of ambidextrous organizations. For justificatory knowledge, we built on business process management (BPM) from a capability perspective and OA antecedents to structure the application domain. This is reasonable for the following arguments: First, maturity models (MM) are valid design products [27] and an established tool for capability development, not only but particularly in the BPM domain [18, 23, 36]. Further, MM intended for a prescriptive purpose of use include good or best practices which is helpful to provide practical guidance [35]. Second, capability development is tightly linked to BPM because capabilities and processes both deal with a coordinated set of tasks and their execution [24, 33, 52]. We therefore rely on BPM to foster OA capability development. Third, focusing on OA antecedents reveals prerequisites for the configuration of ambidextrous organizations and related capabilities [44], whereas outcomes and moderators address the OA-performance linkage providing the rationale for why OA is beneficial. Our OAMM is an initial step offering guidance for OA researchers to systematically develop OA capabilities.

In developing our OAMM, we draw upon the research process for design science as proposed by Peffers et al. [34]. Subsequent to problem identification and motivation as carried out in this introductory section (*research problem*), we deliver on the theoretical background in Sect. 2 and derive design objectives for our problem solution (*objectives for a solution*). Our research approach is presented in Sect. 3. Section 4 is concerned with the design specification of our OAMM (*design and development*) based on the procedure model by Becker et al. [3]. Moreover, our evaluation activities are presented (*demonstration and evaluation*). The conclusion section summarizes the main insights, delivers on both theoretical and practical implications (*communication*), and provides avenues for future research pointing to the limitations of our study.

2 Theoretical Background

2.1 Organizational Ambidexterity

OA is described as an organization's capability to maintain dual capacities for both exploitation and exploration for surviving in dynamic business environments and managing organizational change [51]. Exploitation seeks the refinement of existing products by continuous streamlining of business processes for productivity in operations [42]. Activities related to exploitation are described in terms of efficiency, control, and certainty [19, 26]. Exploration strives for radical innovation of products, services, and processes, to achieve adaptability and growth [42]. Activities related to exploration are associated with experimentation, autonomy, and risk-taking [19, 26].

Considering OA antecedents as one of three major research streams on OA, sequential and simultaneous approaches in implementing OA can be distinguished. Early studies conceptualize OA as the temporal sequencing of exploitation and exploration for their separation over time [17, 39, 43]. In contrast, subsequent studies suggest that tensions between exploitation and exploration do not need to be an either/or proposition and can be addressed simultaneously within the organization [17, 47, 51]. The extant literature concerned with a simultaneous pursuit of OA features three different modes of OA, distinguishing structural, contextual, and leadership-based antecedents [38]. Structural ambidexterity originates from dual organizational structures with independent business units for exploitation and exploration [4, 32]. Contextual ambidexterity anchors the ability to balance exploitation and exploration to individuals [1, 13]. Leadership-based ambidexterity attributes a key role to leadership processes in fostering OA [25, 31]. Thus, we specify the following design objective:

(DO.1) *Ambidextrous Organizations:* To systematically develop OA capabilities, an organization must develop dual capacities for exploitation and exploration. Therefore, sequential and simultaneous approaches, including structural, contextual, and leadership-based antecedents of OA, need to be integrally covered.

2.2 Business Process Management and Capability Development

With process orientation being a central paradigm of organizational design, BPM is closely related to capability development [20, 33]. BPM reflects the skills and routines necessary to integrate, build, and reconfigure an organization's business processes in response to environmental change [12, 50]. Therefore, six core elements of an organization's BPM capability have been identified: strategic alignment, governance, methods, information technology, people, and culture [55]. These elements further split into thirty BPM-related capability areas. Table 1 shows a brief description of the six core elements, for a detailed description see the handbook of BPM [55]. Against the background of dynamic business environments highlighting the importance of OA to sustain success [30], the BPM domain recognizes the need to foster 'ambidextrous BPM' [42]. As such an organization consciously decides whether its BPM should strive for exploitation (e.g., improvement), exploration (e.g., innovation), or both simultaneously. This leads to the following design objective (DO):

(DO.2) *BPM and Capability Development:* To systematically develop an organization's OA capabilities, it is necessary to improve distinct capability areas related to the six core elements of BPM by developing both exploitative and explorative BPM capabilities for each of the core elements (*ambidextrous BPM*).

Table 1. The Six core elements of BPM capability.

Core elements	Description
Strategic alignment	BPM goals and the execution of businesses processes need to be tightly linked to an organization's strategy
Governance	Roles and responsibilities for various levels of BPM need to be appropriately defined for transparent accountability. Governance further relates to designing decision-making and reward processes to guide process-related tasks
Methods	Methods accumulate all tools and techniques that support and enable activities along the process lifecycle and within organization-wide BPM initiatives
Information technology	IT-based solutions such as application and support systems utilized in activities along the process lifecycle and BPM initiatives are comprised within IT
People	People relates BPM capabilities to an organization's human capital and ecosystem. It captures individuals and groups continually enhancing and applying their process skills
Culture	Culture comprises all values and beliefs with respect to an organization built around process orientation. A facilitating environment offers the surrounding for BPM initiatives

2.3 Maturity Models

A vast number of MM have been developed and applied to various domains in the context of BPM [9, 57]. MM are highly appreciated to support organizations in improving their BPM capabilities by elucidating a maturation path along different stages in an anticipated, desired, or logical way [41, 53]. Therefore, MM contain a sequence of maturity stages as well as a descriptions of each stage's characteristics [35]. Progress along the maturation path towards the final state of maturation requires constant improvement related to organizational capabilities [3, 41].

MM serve three purposes of use when practically applied: prescriptive, descriptive, or comparative [3, 35]. A descriptive purpose of use applies if the MM can be used to assess the organization's as-is situation [35]. The MM has a prescriptive purpose of use if it provides guidance on how to determine desirable future maturity stages and suggests initiatives for improvement [35]. A comparative purpose of use is given if the MM serves internal or external benchmarking [35]. To guarantee the usefulness and applicability of MM, first, the process of model design requires substantiation with a procedure model. Second, the model as a design product itself needs to account for design principles [35]. Therefore, the framework of general design principles (DPs) as per Pöppelbuß and Röglinger clusters nine DPs into three nested groups: basic principles, principles for descriptive purpose of use, and principles for prescriptive purpose of use [35]. A depiction of the DPs is provided in Fig. 3 (see Sect. 4). This leads to the following design objective:

(DO.3)   *Maturity Models:* To systematically develop OA capabilities, MM need to be developed following an accepted procedure model and account for general design principles.

3   Research Method

Maturity Model Development. Our study follows the DSR process by Peffers et al. [34] to develop our artifact, i.e., the OAMM that assists organizations in developing OA capabilities based on actionable practices. When formulating the design specification of our OAMM in the design and development phase of the DSR process, we follow the procedure model for MM development by Becker et al. [3] (Fig. 1), supplemented by a literature review and the card sorting approach [59].

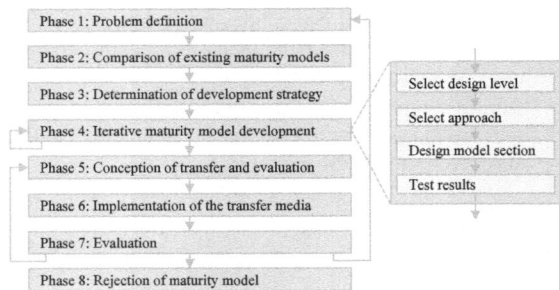

Fig. 1. Procedure model for developing maturity models [2].

Phase 1 to 4 are crucial to develop the design specification of a MM, whereas phases 5 to 8 concern its application and evaluation. Our research comprises the development of the OAMM, while subsequent phases are planned for future research.

The development of our OAMM started from stating the research problem (phase 1) in the introduction. Recognizing a lack of guidance on how to put OA into practice, we address this gap by enhancing the systematic development of OA capabilities. Searching the extant body of knowledge, no MM targeted to OA has been identified (phase 2). Neither CMMI as the archetype of capability MM [7], nor other BPM-related MM [41] are presumed adequate for answering our research question. MM are mostly based on established best practices [7], whereas our OAMM is a first attempt to structure OA capability development with no accurate measures [19]. Thus, we selected a strategy of completely new model design (phase 3). To iteratively develop the OAMM (phase 4), we selected two approaches: (1) literature review to identify required capabilities for OA development, i.e., actionable practices, (2) card sorting approach to assign these practices to different maturity stages. Both approaches are briefly explained below.

Literature Review. We first conducted a literature review [56, 58] to extract actionable practices for OA capability development. We searched Google Scholar [15] and the Web of Science Core Collection [6] such that we assume to have covered core publications from the general OA literature. Using "organizational ambidexterity" as a search term delivered 20,285 results. To obtain a manageable scope of papers, we selected the top 25 search results by number of citations for each of the two databases. In doing so, we assume to cover the most relevant articles which provide us with a sound basis for developing our OAMM as an initial step. Ending up with a list of 50 articles, we first removed duplicates. Second, the relevance of each publication was assessed based on the title and abstract and non-adequate articles were sorted out. We compiled a final list of 15 publications [1, 5, 11, 13, 17, 21, 22, 30–32, 38, 39, 44–46] to be included in our in-depth screening process. Focusing on OA antecedents in screening the articles, we extracted 754 relevant statements and consolidated all that contained the same message. For the remaining statements, we decided if OA antecedents were addressed on a high, medium, or low level of abstraction to exclude all that were not actionable enough to assist organizations in putting OA into practice. To illustrate the three levels, we consider the example of structural ambidexterity: It postulates dual organizational structures [4, 32] (high level), distinguishing mechanistic and organic structures [39] (medium level), which require large and centralized exploitative units and small and decentralized explorative units respectively [38] (low level). We rephrased all remaining statements in a concise and action-oriented manner to become actionable practices. All practices promote OA by either distinguishing clear actions for exploitation and exploration or focus on the ambidextrous idea in general. Thus, there are no practices only addressing exploration or exploitation separately. The result of our literature review provided a set of 44 actionable practices to be included in our OAMM. Structuring our set of practices along the six factors of BPM, we realized that it does not contain practices for IT. However, against the background of digitalization [14], we acknowledge an organization's IT capability as a key component [14, 29]. Consequently, we decided to search for additional articles within the the the AISeL [2] and EJIS [49] databases. Using the search term "IT ambidexterity" and "ambidexterity" within title and abstract leads to 13 articles. We proceeded exactly as we have done before and included three more articles, more precisely 10 more actionable practices.

Card Sorting. After conducting the literature review to identify actionable practices for OA capability development, we used the card sorting approach. Card sorting is generally used to organize and categorize knowledge [59]. It can be performed in an open or closed manner. While in a closed card sorting participants sort content into predetermined categories, an open card sorting asks them to sort and categorize content into their own categories [40]. To ensure reliability of our results, the level of agreement between two raters is calculated [28]. To assess inter-rater reliability, the Cohen's Kappa coefficient is used [8]. It can be interpreted as the proportion of joint judgement in which there is agreement after chance agreement is excluded. In cases of disagreement, the raters discuss all mismatching assessments and decide on one maturity stage.

Evaluation Activities. To evaluate our OAMM, we follow the DSR evaluation framework by Sonnenberg and vom Brocke [48]. Basically, the choice of evaluation strategies occurs along two dimensions: *when* and *how* to evaluate [37, 54]. *When* to carry out the evaluation is determined relative to artifact construction. While ex-ante evaluation happens before the construction of an artifact, ex-post evaluation is conducted afterwards. For *how* the evaluation of an artifact occurs, two different types of evaluation approaches can be distinguished. Artificial approaches refer to the formal proof of an artifact, e.g., by feature comparison, whereas naturalistic approaches cover an initial demonstration by involving real problems, users, tasks and systems, e.g., by case studies. We use an ex-ante artificial evaluation approach, i.e., feature comparison to assess whether the design specification of our OAMM contributes to the solution of our research problem. Therefore, we discuss it against the design objectives derived from justificatory knowledge in Sect. 2.

4 Design Specification of the Maturity Model

4.1 Conceptual Architecture

In this section, we provide insights into the *design and development* of our OAMM (phase 4) by presenting the results of conducting the four sub-steps of this phase: selecting the design level, selecting the approach, designing the model section, and testing the results (Fig. 1). Our OAMM is intended to provide guidance for the structured development of an organization's OA capabilities. Therefore, we present our OAMM as a matrix, where the vertical axis includes the six core elements of BPM, corresponding capability areas, and underlying actionable practices as criteria for maturity assessment. The horizontal axis includes five consecutive maturity stages (Table 2). To compile the overall architecture of our OAMM, we performed a closed card sorting assigning each actionable practice to exactly one predefined maturity stage.

For the vertical axis, we structure our set of 54 actionable practices along the six core elements of BPM and corresponding capability areas, which have already been appreciated by researchers across various domains [55]. This seems reasonable as capability development is tightly linked to BPM because capabilities and processes both deal with a coordinated set of tasks and their execution [33, 52]. The six core elements of BPM are further presumed to provide a comprehensive description of all areas of organizational design which embody an organization's BPM capabilities. Moreover, there is no alternative classification that we considered to fit our research.

For the horizontal axis, we derived five maturity stages based on the Dreyfus model of directed skill acquisition which describes developmental stages for how individuals acquire skills [10]. The model reveals progressive changes in a performer's perception of their task environment assuming that advanced skills lead to less dependency on abstract principles or instruction and more on concrete experience [10]. We suggest a fit

Table 2. OAMM maturity stages and stage characteristics.

Maturity stages	Stage characteristics
(1) Novice	The novice organization is given instructions for acting based on objectively defined rules. These rules are independent of the OA domain and can be understood without OA capabilities. The organization lacks ambidextrous thinking and behavior. It is indifferent towards ambidextrous strategies and related outcomes
(2) Advanced beginner	The advanced beginner organization gains understanding of the OA domain. The organization has some experience coping with real cases. Specific requirements in pursuit of ambidextrous strategies are recognized. The organization is indifferent towards related outcomes as ambidextrous thinking is not disseminated
(3) Competent	The competent organization perceives multiple antecedents and requirements of OA and judges on their relative importance based on instruction or experience. It strives for routines in showing ambidextrous behavior. The organization recognizes ambidextrous goals, but does not take on ambidextrous attitudes. It is concerned with the positive or negative consequences of ambidextrous strategies
(4) Proficient	The proficient organization is aware of ambidextrous goals. Requirements related to ambidextrous goals and behaviors can be prioritized with respect to specific situations. A holistic view enables the organization to intuitively recognize challenges and benefits of OA. It still needs rules for action and guidance on how to put OA into practice. Ambidextrous thinking and attitudes are demonstrated
(5) Expert	The expert organization draws on substantial experience in the OA domain. Dual capacities for exploitation and exploration enable immediate situational responses. Knowing which reaction is best to accomplish a certain goal, decision making and allocation of resources to exploitation and exploration are based on intuitive expertise. The organization is fully committed to the pursuit of ambidextrous strategies

between the model and our research goal of enhancing structured OA capability development as the development of an organization's capabilities can be tightly linked to learning patterns and skill development of individuals [11]. We labelled our OAMM maturity stages in accordance with the skill levels contained in the Dreyfus model: (1) novice, (2) advanced beginner, (3) competent, (4) proficient, (5) expert. In contrast, the stages' definitions have been adapted by retaining general definitory elements from the Dreyfus model and respecting characteristics of the OA domain within our definition. Table 2 depicts the five maturity stages as contained in our OAMM.

4.2   Card Sorting and Final Results

To compile the overall architecture of our OAMM, we performed a closed card sorting. Two authors were provided with the identified set of 54 actionable practices and asked to independently assign each practice to one maturity stage. Our OAMM as the resulting artifact is shown in Fig. 2. Thus, the assignment of each practice to one maturity stage as well as the percentage of practices associated with each of the six core elements and maturity stages is given. Based on these card sorting results, the inter-rater reliability was calculated using Cohen's Kappa [8]. We achieved a value of 0.67, which indicates reliability of our results [28].

Finally, testing the results of compiling our OAMM, we first present key findings with respect to the six core elements, its capability areas, and actionable practices, i.e., vertical axis. Thereby, we also account for maturation paths which can be seen as sequences of actionable practices related to a distinct capability area. Second, we discuss key findings with respect to maturity stages, i.e., horizontal axis. Third, we tested our results for comprehensiveness, consistency, and problem adequacy [3]. It is worth mentioning that all key findings reflect particularities of the sample reviewed for purposes of our study and therefore, the distribution of practices per core elements and maturity stages as well as all related insights are highly dependent on our research approach.

Vertical Axis. As for the six core elements, our OAMM covers all factors. People comprises around one fourth of all practices, followed by strategic alignment, IT, governance, culture, and methods. This distribution is reasonable as it resembles the relative importance of different OA antecedents as presented in the existing body of knowledge. For example, the pivotal role of the top management team in balancing exploitation and exploration is recognized [47]. OA capabilities related to leadership skills and behaviors are comprised within the *people* dimension, suggesting its strong presence in our OAMM. Besides the leadership-based approach, the literature is largely concerned with structural antecedents of OA [21], pointing to the relative importance of *strategic alignment* as revealed in our OAMM. Further, our OAMM reveals that the development of an organization's ambidextrous IT capabilities is a strategic issue. Investments in digital technologies need to be cautiously orchestrated to align with existing IT capabilities complementary IT portfolios [29] and avoid excessive costs for resource integration.

Additionally, fourteen maturation paths within various capability areas could be identified. We consider two illustrative examples. First, the capability area 'roles and responsibilities' contains four practices. The related maturation path outlines their desired implementation order as indicated by consecutive maturity stages (2) to (5). Organizations systematically develop OA capabilities by implementing the practice located with maturity stage (2) first and stepwise completing practices along the maturation path. Second, the capability area 'enterprise process architecture' contains five practices. The two practices relating to sequential approaches are located at maturity stages (2) and (3) and thus precede those three practices relating to simulta-neous approaches and located at maturity stages (3) to (5). This finding complies with the consecutive emergence of sequential and simultaneous approaches in the literature [17, 39].

Horizontal Axis. Analyzing the number of practices per maturity stages provides some interesting insights. Only two practices have been associated with maturity stage (1). These practices reflect general requirements conducive to OA, but need to be implemented independent of domain-specific characteristics. Moreover, while all practices feature the ambidextrous idea, novice organizations show only rare or no properties of OA at all. A majority of 34 practices has been assigned to maturity stages

Capability Areas		Ref.*	Actionable Practices	Maturity Stages				
				(1)	(2)	(3)	(4)	(5)
Strategic Alignment (25%) — Strategy & Process / Capability Linkage		[33]	Pursue further growth in a single core business and a limited expansion around that core into closely related new areas.	X				
		[33]	Adopt a combination of induced strategy processes (exploitation) and autonomous strategy processes (exploration).		X			
		[18][39][8]	Externalize either exploitation or exploration to achieve efficient specialization across an inter-organizational network, e.g. strategic alliances, corporate ventures, spin-off entities.		X			
		[13]	Establish both additional relationships with existing partners (exploitation) and relationships with new partners (exploration) when externalizing either exploitation or exploration in an inter-organizational network.			X		
		[38]	Strive for diversity and good connections but not the utmost central position in a network, to make the externalization of either exploitation or exploration effective by realizing information and resource benefits as well as strategic integration.				X	
		[18][34][33]	Foster the ability to internally apply knowledge, which has been accessed outside the organization by externalization of either exploitation or exploration, through own R&D efforts.					X
Enterprise Architecture / Process Architecture		[27][34][33]	Use semi- or quasi-structures to shift formal organization structures over time, for the dynamic and temporal sequencing of exploitation and exploration.	X				
		[39][13][18]	Develop switching rules and appropriate change routines to facilitate transitions in the temporal sequencing of exploitation and exploration.			X		
		[34][27][18]	Deploy dual structural arrangements in a simultaneous pursuit of exploitation and exploration, creating large and centralized exploitative units with mechanistic structures, and small and decentralized exploitative units with organic structures.			X		
		[34][10][39]	Create team-based structures to pursue exploitation and exploration simultaneously within one single business unit, whereby one group adopts a mechanistic structure while another takes on an organic structure.			X		
		[34][33]	Build parallel organizational structures, e.g. quality circles, that enable people from the same unit to switch between the mechanistic structure for exploitation and organic structures for exploration.					X
Governance (20%) — Decision Making / Mgmt Process		[10][26][17]	Impose top-down direction for definitive resource allocation decisions, while creating mechanisms that allow actors at lower hierarchical levels to access the resources available to others.			X		
		[33][34][40]	Involve managers with dynamic decision making, such that they repeatedly and intentionally orchestrate firm resources.				X	
Process Roles and Responsibilities		[27][10][11]	Attribute clear job descriptions and instructions to individuals in business units focused on either exploitation or exploration, for explicate roles and segregated work modes, i.e. routine and non-routine responsibilities.	X				
		[13][11]	Employ distinct roles, such that extrinsically motivated individuals perform tasks focused on acting appropriately (exploitation), and those intrinsically motivated take on activities focused on creativity (exploration).			X		
		[11][17][18]	Switch roles and responsibilities by varying the nature of work at different times rather than addressing exploration and exploitation simultaneously, e.g. by change of project focus and project rotation.				X	
		[11][10][27]	Adopt routines to systematize the creative process, such that employees follow well-defined processes and standardized best practices for exploitation, while continuously having possibilities of experimentation for exploration.					X
Process Related Standards		[25][33][18]	Establish a top management incentive system that makes individual benefits dependent on a team's outcome and the overall firm performance, such that no individual agendas are pursued.			X		
		[10][38][18]	Adopt clear objectives and goal-setting programs for distinct standards of performance (exploitation), while setting aggressive but not unrealistic targets to encourage individuals to push for ambition goals (exploration).				X	
		[11][10][38]	Value both exploitation and exploration activities and reinforce them with reward and recognition, while being consistent in the application of sanctions.				X	
Methods (7%) — Program & Project Management Process		[11][26][18]	Diversify your project portfolio by selecting routine projects for exploitation as well as high-risk projects for exploration.		X			
		[1]	Iterate between work modes of project control (exploitation) and freedom in projects (exploration) for their temporal separation, and increase iterations in frequency as projects progress.				X	
		[1]	Foster improvisation in projects, such that project work follow clear processes and adheres to defined goals (exploitation), while simultaneously facilitating creative expression to move beyond customer constraints (exploration).					X

(%) Distribution of core elements / maturity stages by number of associated practices; OAMM Maturity Stages: (1) = Novice, (2) = Advanced Beginner, (3) = Competent, (4) = Proficient, (5) = Expert. A detailed list of references will be provided upon request.

*Note: We listed a maximum of three references with each practice for reasons of clarity.

□ Assigned maturity stage — Maturation path

Fig. 2. Organizational ambidexterity maturity model (OAMM).

Six Core Elements		Capability Areas	References	Actionable Practices	Maturity Stages				
					(1)	(2)	(3)	(4)	(5)
People (32%)		Education Process	[1][10]	Establish socialization practices – from hiring to ongoing reviews – to help employees identify themselves as paradoxical work identities with both discipline (exploitation) and passion (exploration).				x	
			[10][39]	Provide job enrichment programs for education, training and experience in both exploitation and exploration.			x		
			[10][13][33]	Instrumentalize decentralized and direct interpersonal learning for implicit knowledge transfer, e.g. sharing best practices.					x
		Collaboration Process	[40][38][10]	Facilitate open discussion about tensions between exploitation and exploration to foster acceptance.			x		
			[17][34]	Establish cross-functional interfaces at lower hierarchical levels, e.g. cross-functional teams, projects, temporary workgroups, task forces, liaison personnel.			x		
			[34][17][33]	Acquire high levels of top-down, bottom-up, and horizontal knowledge flows for connectedness to all hierarchical levels and integration of differentiated efforts for exploitation and exploration.				x	
			[11][34][33]	Create interactions across different organizational levels to cope with exploitation and exploration on multiple levels and leverage synergies, e.g. between business unit, group, individual level				x	
		Leaders Management Process	[13][25][33]	Compose top management teams that are heterogeneous, regarding its members' diversity of hierarchical status, knowledge and experience working together, e.g. adopting a mix between "newcomers" and "old-timers"	x				
			[18]	Enable risk-averse decision makers to drive exploitation, and draw on risk-prone managers for purposes of exploration.		x			
			[17][27][39]	Assign different leadership behaviors and management styles, such that transactional leadership is related to exploitation, whereas transformational leadership is related to exploration.		x			
			[10][40]	Adopt management strategies of acceptance rather than defensiveness towards exploitation-exploration tensions in combination with resolution strategies.			x		
			[25][38][34]	Develop conflict resolution skills of the top management team to openly discuss conflicting task issues and freely exchange differing knowledge for integration of exploitation and exploration.			x		
			[18][33][38]	Develop coordination skills of the top management team related to collaborative behavior, information exchange, and joint decision making, for a common agenda and integration of exploitation and exploration.				x	
			[10][39][40]	Employ effective leaders with a breadth of past experience who have the cognitive capabilities and behavioral repertoires to engage in paradoxical thinking and take on complex tasks related to ambidextrous management.					x
Culture (16%)		Values & Beliefs Process	[27][25][39]	Develop an overarching strategic intent and a common paradoxical vision.		x			
			[25][26][11]	Relentlessly communicate the ambidextrous strategy and reiterate each supportive communication, to infuse the paradoxical vision and reinforce dual purposes for exploitation and exploration.				x	
		Leader-ship Attention Process	[25][26][10]	Make the pursuit of an ambidextrous strategy compelling and demote strategic coherence through shared ambitions and a collective identity					x
			[18][10][33]	Ensure support of top management for the organizational restructuring, that modes of temporal or structural differentiation between exploitation and exploration entail.			x		
		Behaviours & Attitudes Process	[34][25][10]	Employ managers that engage in both exploitation and exploration activities, and especially act as pioneers in triggering creativity for exploration.				x	
			[10][27][38]	Nurture a flexible organization culture that equally builds on performance management and social support, striving for a balance between norms for control (exploitation) and collaboration (exploration)			x		
			[8][40][5]	Foster cognitive and behavioral complexity as well as calmness of your employees, for acceptance of exploitation-exploration tensions and adoption of appropriate behavior to allocate their time accordingly.					x
		Assigned maturity stage		(%) Distribution of core elements / maturity stages by number of associated practices; OAMM Maturity Stages: (1) = Novice, (2) = Advanced Beginner, (3) = Competent, (4) = Proficient, (5) = Expert. A detailed list of references will be provided upon request.	5%	14%	34%	27%	20%

* Note: We listed a maximum of three references with each practice for reasons of clarity. A detailed list of references will be provided upon request.

☐ Maturation path

Fig. 2. (*continued*)

(3) and (4). This can be explained as it is easier to develop capabilities on lower maturity stages, while it is more difficult to finally reach the highest maturity stage (5). Additionally, searching the specific OA literature is most likely to address advanced OA capabilities and thus reveal practices located at the higher maturity stages, while the general literature as covered by our study is presumed to address more basic requirements of lower stages. Another interesting insight show that initial stages in maturation of the ambidextrous organization require the accomplishment of practices associated with strategic alignment and people, whereas for example the demonstration of ambidextrous IT capabilities mostly requires proficient (4) or expert (5) stage.

Test for Comprehensiveness, Consistency, and Problem Adequacy. Overall, we assume comprehensiveness of maturity assessment based on the OAMM as we built on an established framework of BPM capabilities [55]. Yet we acknowledge that reducing the overall number of hits from our literature review (see Sect. 3) limits the comprehensiveness of our set of practices. A satisfactory level of inter-rater reliability for our card sorting indicates consistency of our results. We further postulate problem adequacy as our OAMM contains various maturation paths, which supports our goal of enhancing prescriptive knowledge on OA capability development.

5 Evaluation

In line with our evaluation strategy based on the DSR evaluation framework by Sonnenberg and vom Brocke [48], we conduct an artificial ex-ante evaluation by discussing the design specification of our OAMM against the three design objectives derived in Sect. 2. Figure 3 shows the results of our feature comparison. In sum, feature comparison revealed that our OAMM address all three design objectives, but not to the full extent. The OAMM is beset with some limitations from a theoretical perspective for the sake of increased applicability. We capture the resulting need for future research in the conclusion.

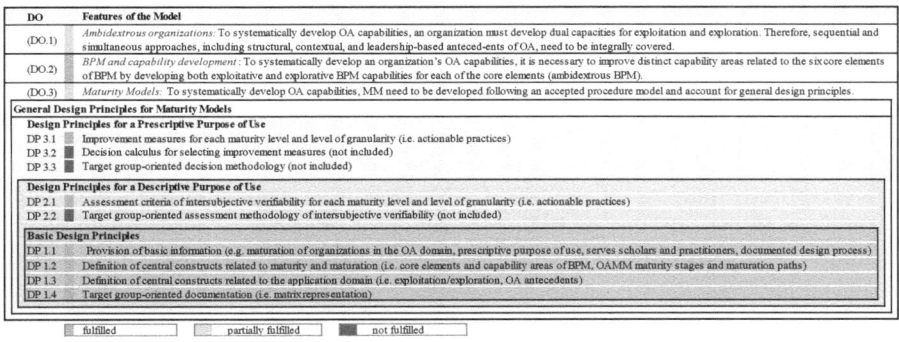

Fig. 3. Results of feature comparison.

6 Conclusion

To thrive in turbulent competitive environments, it is vital for organizations to develop OA capabilities, i.e., dual capacities for exploitation and exploration [38, 51]. Following an identified need for guidance on *how to put OA into practice* [1, 31], we developed our OAMM to assist organizations in acquiring OA capabilities. To do so, we built on the six core elements of BPM and corresponding capability areas [55] as well as five maturity stages to anticipate fourteen related maturation paths based on 44 actionable practices. Our OAMM is meant to serve as a starting point for structured OA capability development and paves the way for maturation towards an ambidextrous organization.

The results of our research have implications for both academia and practice. As for theoretical implications, first, our set of actionable practices consolidates insights from different research streams related to OA antecedents. Therefore, our literature review identifies OA capabilities for different organizational levels, i.e., the corporate, business unit, group, and individual level. This multi-level concept is important to fully capture an organization's exploitation and exploration activities [38]. Second, the architecture of our OAMM provides insights into OA from a BPM perspective. This is done by applying the six core elements of BPM and related capability areas to structure our set of actionable practices derived from OA literature. Additionally, we outline maturation paths for organizations based on various capability areas to advance from novice (1) to expert stage (5) regarding ambidextrous capabilities. Finally, our OAMM lays the groundwork for further elaboration, i.e., including additional practices and deriving further maturation paths. Therefore, our OAMM facilitates the classification of actionable practice along the six core elements of BPM and its capability areas [55] as well as five maturity stages (Table 2). In sum, our OAMM is an initial step towards a MM offering guidance to systematically develop OA capabilities. Our OAMM thereby contributes to prescriptive knowledge in respect of *how to put OA into practice*, i.e., on how the desired balance between exploitation and exploration can be achieved. Future OA research can use our OAMM to identify and structure further antecedents based on the six factors. If further antecedents are identified and analyzed in a conceptual way, these insights can be subsequently integrated in the OAMM to facilitate the development of an ambidextrous organization.

As for practical implications, our OAMM assists practitioners in implementing OA and configuring ambidextrous organizations. Delivering on a descriptive purpose of use, our OAMM allows organizations to assess their as-is-situation and provides a sufficient basis for determining an organization's current state of maturity. Delivering on a prescriptive purpose of use, our OAMM outlines maturation paths and guides practitioners in the selection and implementation of practices associated with distinct capability areas. Thereby, our set of actionable practices lays the foundation for enhancing the development of OA capabilities. It is noteworthy that practitioners may alter maturation paths in respect of their organization's specific situation, prioritizing the development of certain capability areas and implementing practices along the respective maturation paths until a satisfactory level of OA capabilities is reached.

Pointing to the limitations of our study, we present several avenues for future research. First, our findings build on a selection of 15 articles from the general OA literature. Thereby, we focused on the most cited studies since they cover core OA research. As a consequence, more specific fields of OA capabilities, e.g., capabilities for information technology or methods, are partly covered. Consequently, DO.1 is only partly fulfilled. Further research should include a more extensive literature search by searching more databases and including specific OA literature, e.g., literature investigating IT ambidexterity or exploitation and exploration methods. Second, we performed the card sorting from a researchers' perspective only. We believe, however, that the current card sorting is adequate to provide first insights on the development of OA capabilities based on various maturation paths. Further research may perform the card sorting with both researchers and practitioners. Third, in developing our OAMM we followed Becker et al.'s [3] procedure model, but did not finalize the whole procedure. We performed all crucial steps through phase 4 and then moved on to the evaluation of our OAMM (phase 7). To complete the model development procedure, a transfer to academics and practitioners is suggested. Moreover, the evaluation (phase 7) should be extended to assess the applicability and usefulness of our OAMM in naturalistic settings, e.g., conducting expert interviews or real-world case studies. Fourth, the artificial ex-ante evaluation of our OAMM revealed that general design principles for MM are addressed, but not to the full extent. To fully serve the intended prescriptive purpose of use, we suggest developing our OAMM further as a stand-alone artifact by including a decision calculus for the selection of improvement measures and some target group-oriented decision methodology. Overall, we call for future research in the area of structured OA capability development to address the currently observed imbalances in the number of practices assigned to core elements and maturity stages which reflect the particularities of this study's research approach.

References

1. Andriopoulos, C., Lewis, M.W.: Exploitation-exploration tensions and organizational ambidexterity. Managing paradoxes of innovation. Organ. Sci. **20**(4), 696–717 (2009)
2. Association for Information Systems: Association for Information Systems Electronic Library (AISeL). http://aisel.aisnet.org. Accessed 25 May 2018
3. Becker, J., Knackstedt, R., Pöppelbuß, J.: Developing maturity models for IT management. Bus. Inf. Syst. Eng. **1**(3), 213–222 (2009)
4. Benner, M.J., Tushman, M.L.: Exploitation, exploration, and process management. The productivity dilemma revisited. Acad. Manag. Rev. **28**(2), 238–256 (2003)
5. Cao, Q., Gedajlovic, E., Zhang, H.: Unpacking organizational ambidexterity. Dimensions, contingencies, and synergistic effects. Organ. Sci. **20**(4), 781–796 (2009)
6. Clarivate Analytics: Web of Science. www.webofknowledge.com. Accessed 09 Jan 2018
7. CMMI Product Team: CMMI® for Development, Version 1.3 (No. CMU/SEI-2010-TR-033). Improving Processes for Developing Better Products and Services. Software Engineering Institute (2010)
8. Cohen, J.: A coefficient of agreement for nominal scales. Educ. Psychol. Meas. **20**(1), 37–46 (1960)

9. de Bruin, T., Freeze, R., Kaulkarni, U., Rosemann, M.: Understanding the main phases of developing a maturity assessment model. In: Proceedings of the 16th Australasian Conference on Information Systems (ACIS), Sydney, NSW, Australia, Paper 109 (2005)

10. Dreyfus, S.E., Dreyfus, H.L.: A Five-Stage Model of the Mental Activities Involved in Directed Skill Acquisition. California Univ Berkeley Operations Research Center (1980)

11. Eisenhardt, K.M., Furr, N.R., Bingham, C.B.: Microfoundations of performance. Balancing efficiency and flexibility in dynamic environments. Organ. Sci. 21(6), 1263–1273 (2010)

12. Forstner, E., Kamprath, N., Röglinger, M.: Capability development with process maturity models. Decision framework and economic analysis. J. Decis. Syst. 23(2), 127–150 (2014)

13. Gibson, C.B., Birkinshaw, J.: The antecedents, consequences, and mediating role of organizational ambidexterity. Acad. Manag. J. 47(2), 209–226 (2004)

14. Gimpel, H., Röglinger, M.: Digital Transformation: Changes and Chances. Insights Based on an Empirical Study. http://www.fim-rc.de/expertise/digitalization/?lang=en. Accessed 10 May 2018

15. Google LLC: Google Scholar Database. https://scholar.google.de. Accessed 09 Sept 2018

16. Gregor, S., Hevner, A.R.: Positioning and presenting design science research for maximum impact. MIS Q. 37(2), 337–366 (2013)

17. Gupta, A.K., Smith, K.G., Shalley, C.E.: The interplay between exploration and exploitation. Acad. Manag. J. 49(4), 693–706 (2006)

18. Harmon, P.: Process Maturity Models. http://www.bptrends.com/publicationfiles/spotlight_051909.pdf. Accessed 09 Mar 2018

19. He, Z.-L., Wong, P.-K.: Exploration vs. exploitation. An empirical test of the ambidexterity hypothesis. Organ. Sci. 15(4), 481–494 (2004)

20. Helfat, C.E., Peteraf, M.A.: The dynamic resource-based view. Capability lifecycles. Strat. Manag. J. 24(10), 997–1010 (2003)

21. Jansen, J.J.P., Tempelaar, M.P., van den Bosch, F.A.J., Volberda, H.W.: Structural differentiation and ambidexterity. The mediating role of integration mechanisms. Organ. Sci. 20(4), 797–811 (2009)

22. Lavie, D., Stettner, U., Tushman, M.L.: Exploration and exploitation within and across organizations. Acad. Manag. Ann. 4(1), 109–155 (2010)

23. Lehnert, M., Linhart, A., Röglinger, M.: Chopping down trees vs. sharpening the axe. Balancing the development of BPM capabilities with process improvement. In: Sadiq, S., Soffer, P., Völzer, H. (eds.) BPM 2014. LNCS, vol. 8659, pp. 151–167. Springer, Cham (2014). https://doi.org/10.1007/978-3-319-10172-9_10

24. Lehnert, M., Linhart, A., Röglinger, M.: Value-based process project portfolio management. Integrated planning of BPM capability development and process improvement. Bus. Res. 9(2), 377–419 (2016)

25. Lubatkin, M.H., Simsek, Z., Ling, Y., Veiga, J.F.: Ambidexterity and performance in small- to medium-sized firms. The pivotal role of top management team behavioral integration. J. Manag. 32(5), 646–672 (2006)

26. March, J.G.: Exploration and exploitation in organizational learning. Organ. Sci. 2(1), 71–87 (1991)

27. March, S.T., Smith, G.F.: Design and natural science research on information technology. Decis. Support Syst. 15(4), 251–266 (1995)

28. Nahm, A.Y., Rao, S.S., Solis-Galvan, L.E., Ragu-Nathan, T.S.: The Q-sort method. Assessing reliability and construct validity of questionnaire items at a pre-testing stage. J. Mod. Appl. Stat. Methods 1(1), 114–125 (2002)

29. Nwankpa, J.K., Datta, P.: Balancing exploration and exploitation of IT resources. The influence of digital business intensity on perceived organizational performance. Eur. J. Inf. Syst. (EJIS) 26(5), 469–488 (2017)

30. O'Reilly, C.A., Tushman, M.L.: Ambidexterity as a dynamic capability. Resolving the innovator's dilemma. Res. Organ. Behav. **28**, 185–206 (2008)
31. O'Reilly, C.A., Tushman, M.L.: Organizational ambidexterity in action. How managers explore and exploit. Calif. Manag. Rev. **53**(4), 5–22 (2011)
32. O'Reilly, C.A., Tushman, M.L.: Organizational ambidexterity. past, present, and future. Acad. Manag. Perspect. **27**(4), 324–338 (2013)
33. Ortbach, K., Plattfaut, R., Pöppelbuß, J., Niehaves, B.: A dynamic capability-based framework for business process management. Theorizing and empirical application. In: Proceedings of the 45th Hawaii International Conference on System Sciences, Maui, Hawaii, pp. 4287–4296 (2012)
34. Peffers, K., Tuunanen, T., Rothenberger, M.A., Chatterjee, S.: A design science research methodology for information systems research. J. Manag. Inf. Syst. **24**(3), 45–77 (2008)
35. Pöppelbuß, J., Röglinger, M.: What makes a useful maturity model? A framework of general design principles for maturity models and its demonstration in business process management. In: Proceedings of the 19th European Conference on Information Systems (ECIS), Paper 28 (2011)
36. Pöppelbuß, J., Niehaves, B., Simons, A., Becker, J.: Maturity models in information systems research. Literature search and analysis. Commun. Assoc. Inf. Syst. (CAIS) **29**(1), 1–15 (2011)
37. Pries-Heje, J., Baskerville, R., Venable, J.: Strategies for design science research evaluation. In: Proceedings of the 16th European Conference on Information Systems (ECIS), Galway, Ireland, Paper 87 (2008)
38. Raisch, S., Birkinshaw, J.: Organizational ambidexterity. Antecedents, outcomes, and moderators. J. Manag. **34**(3), 375–409 (2008)
39. Raisch, S., Birkinshaw, J., Probst, G., Tushman, M.L.: Organizational ambidexterity. Balancing exploitation and exploration for sustained performance. Organ. Sci. **20**(4), 685–695 (2009)
40. Righi, C., James, J., Beasley, M., Day, D.L., Fox, J.E., Gieber, J., Howe, C., Ruby, L.: Card sort analysis best practices. J. Usabil. Stud. **8**(3), 69–89 (2013)
41. Röglinger, M., Pöppelbuß, J., Becker, J.: Maturity models in business process management. Bus. Process Manag. J. **18**(2), 328–346 (2012)
42. Rosemann, M.: Proposals for future BPM research directions. In: Ouyang, C., Jung, J.-Y. (eds.) AP-BPM 2014. LNBIP, vol. 181, pp. 1–15. Springer, Cham (2014). https://doi.org/10.1007/978-3-319-08222-6_1
43. Siggelkow, N., Levinthal, D.: Temporarily divide to conquer. centralized, decentralized, and reintegrated organizational approaches to exploration and adaptation. Organ. Sci. **14**(6), 650–669 (2003)
44. Simsek, Z.: Organizational ambidexterity. Towards a multilevel understanding. J. Manag. Stud. **46**(4), 597–624 (2009)
45. Simsek, Z., Heavey, C., Veiga, J.F., Souder, D.: A typology for aligning organizational ambidexterity's conceptualizations, antecedents, and outcomes. J. Manag. Stud. **46**(5), 864–894 (2009)
46. Smith, W.K., Lewis, M.W.: Toward a theory of paradox. A dynamic equilibrium model of organizing. Acad. Manag. Rev. **36**(2), 381–403 (2011)
47. Smith, W.K., Tushman, M.L.: Managing strategic contradictions. A top management model for managing innovation streams. Organ. Sci. **16**(5), 522–536 (2005)
48. Sonnenberg, C., vom Brocke, J.: Evaluation patterns for design science research artefacts. In: Helfert, M., Donnellan, B. (eds.) EDSS 2011. CCIS, vol. 286, pp. 71–83. Springer, Heidelberg (2012). https://doi.org/10.1007/978-3-642-33681-2_7

49. Taylor & Francis Online: European Journal of Information Systems (EJIS). https://www. tandfonline.com/toc/tjis20/current. Accessed 25 May 2018

50. Teece, D., Pisano, G., Shuen, A.: Dynamic capabilities and strategic management. Strat. Manag. J. **18**(7), 509–533 (1997)

51. Tushman, M.L., O'Reilly, C.A.: Ambidextrous organizations. Managing evolutionary and revolutionary change. Calif. Manag. Rev. **38**(4), 8–29 (1996)

52. van Looy, A., de Backer, M., Poels, G.: Defining business process maturity. A journey towards excellence. Total Qual. Manag. Bus. Excell. **22**(11), 1119–1137 (2011)

53. van Looy, A., Poels, G., Snoeck, M.: Evaluating business process maturity models. J. Assoc. Inf. Syst. **18**(6), 461–486 (2017)

54. Venable, J., Pries-Heje, J., Baskerville, R.: A comprehensive framework for evaluation in design science research. In: Peffers, K., Rothenberger, M., Kuechler, B. (eds.) DESRIST 2012. LNCS, vol. 7286, pp. 423–438. Springer, Heidelberg (2012). https://doi.org/10.1007/978-3-642-29863-9_31

55. vom Brocke, J., Rosemann, M.: The six core elements of business process management. In: vom Brocke, J., Rosemann, M. (eds.) Handbook on Business Process Management 1. Introduction, Methods, and Information Systems, pp. 107–122. Springer, Heidelberg (2010). https://doi.org/10.1007/978-3-642-00416-2_5

56. vom Brocke, J., Simons, A., Niehaves, B., Niehaves, B., Reimer, K., Plattfaut, R., Cleven, A.: Reconstructing the giant. On the importance of rigour in documenting the literature search process. In: Proceedings of the 17th European Conference on Information Systems (ECIS), Verona, Italy, Paper 161 (2009)

57. Weber, C., Curtis, B., Gardiner, T.: Business Process Maturity Model (BPMM), Version 1.0. https://www.omg.org/spec/BPMM/1.0/PDF. Accessed 09 Feb 2018

58. Webster, J., Watson, R.: Analyzing the past to prepare for the future. Writing a literature review. MIS Q. **26**(2), 13–23 (2002)

59. Wood, J.R., Wood, L.E.: Card sorting. current practices and beyond. J. Usabil. Stud. **4**(1), 1–6 (2008)

A Data Governance Framework for Platform Ecosystem Process Management

Sung Une Lee[1,2(✉)], Liming Zhu[1,2], and Ross Jeffery[1,2]

[1] Architecture and Analytics Platforms Group,
Data61, CSIRO, Sydney, Australia
{Sungune.Lee, Liming.Zhu, Ross.Jeffery}@data61.csiro.au
[2] School of Computer Science and Engineering, UNSW, Sydney, Australia

Abstract. Platform ecosystem today is regarded as the key business concept of organizations to win market. Platform companies can grow fast through the data contribution of multi-sided networks. Yet, they face difficulties in managing the data resulted from complicated contribution, use and interactions between the multiple parties. The circumstance causes serious concerns about unclear data ownership and invisible use of data, and ultimately leads to data abuse/misuse or privacy violation. To alleviate to this, a particular type of data governance is required. However, there is limited research on data and data governance for platform ecosystems. We introduce a new data governance framework for platform ecosystems which consists of data, role, decisions and due processes. The framework supports organizations in understanding to show how the risks should be dealt in the processes for business success. We compare 19 existing industry governance frameworks and academic work with our framework to show current gaps and limitations.

Keywords: Data governance · Platform ecosystem · Business process

1 Introduction

All organizations use and manage data. Traditional business companies focus on the management of the "ilities" (availability or usability) of enterprise data. Since the concept of platform ecosystem (PE or business ecosystem) has been widely spread, many organizations can facilitate reaching critical mass by data contribution of two or more external sides. The collected data is analyzed or shared to add value to the companies. This generates more data and it is used by the partners or family companies and the platform users. A negative externality arises from the fact that there are complicated interactions between multiple parties providing, using or sharing data. There is a pervasive problem of data breach (data abuse, misuse or privacy issue) in business ecosystem area [1, 2]. A platform owner company should impose certain regulations on the user participation to reap the benefits of ecosystem growth [3]. Lack or poor implementation of governance causes significant destructive effects on business success.

Data governance refers to comprehensive control, including processes, policies and structures about data asset. It enables a platform owner to orchestrate the complicated processes and relationships affected by multiple parties' participation [4]. In traditional

© Springer Nature Switzerland AG 2018
M. Weske et al. (Eds.): BPM Forum 2018, LNBIP 329, pp. 211–227, 2018.
https://doi.org/10.1007/978-3-319-98651-7_13

data governance, data ownership is clear and simple as there is limited use of data and interactions within an enterprise. Researchers articulate a set of concerns such as unclear data ownership or invisible data supply chain in PEs [5–8]. It must be addressed in data governance to win market. In particular, the role of due process has been highlighted by researchers [8]. However, previous platform studies pay little attention to the importance and the role of data [9]. It leads to limited research in understanding how organizations should manage business processes differently for PE.

We aim to provide data governance for enterprises which run PE business. We identified data governance factors for PEs as a starting point [10]. In this study, we provide a comprehensive data governance framework which comprises three core elements, and discusses current issues and how to improve business capability. Based on that, we suggest due processes as a supporting element of the framework. The due processes encourage desirable behavior of all participating groups to mitigate business risks. This article delivers broad information and knowledge of PE and data governance through a survey on industry platforms and literature review. It helps researchers and practitioners to comprehend how data governance processes should be managed and implemented, and to plan next steps.

The next section presents background information to help an overall understanding of the general concept of data governance and PE, and the current state of academic works on platform governance including our previous studies. Section 3 describes the methodology of this study. Section 4 introduces a data governance framework for PEs. We present each element of the framework: platform data, role and decision domains, and discuss current issues and the possible solutions. We then illustrate due processes along with the data management flow of a PE. In Sect. 5, 19 industry governance frameworks and academic works are reviewed and compared with our framework. We then conclude this study in Sect. 6.

2   Background

Data governance determines who holds the decision rights and is held accountable for decision-making about data assets [11]. To support right decision-making and encourage desirable behavior, it provides comprehensive control such as processes, policies and structures. Khatri and Brown [11] noted data governance decision domains and showed how the domains align with those of IT governance. Weber et al. [12] focused on a context-based approach for data governance design by presenting how organizational contingencies influence on data governance. Those studies, however, are focused on the general business context of organizations where there are simple interactions and internal considerations in using and managing data.

PEs provide a meeting place, and facilitate interactions between two (or more) participating groups [13]. Smedlund and Faghankhani [4] noted traditional organizations easily control participants (employees) and the relationship between them, but platform owners have limited power and ability to fully control platforms as there are multiple parties contributing, deriving and using data. Governance for PEs thus should deliberate the different business context and concepts. Trust, roles, revenue sharing and control are identified as fundamental governance concepts for organizations which run a PE [9, 14, 15]. Those concepts should be implemented in data governance of PEs to encourage good practices of governance and to create value in the use of data [12]. Prior studies on platform governance largely neglect the role of data and data governance, and therefore data governance studies have been rarely found.

In our previous studies [10], we surveyed four platform companies (Facebook, YouTube, EBay and Uber) to show the state of practice of data governance. It revealed the fact that the policies of the platforms are imprecise in terms of data ownership and data usage. It can cause uncertainties and arguments between participating groups and business risks. We also reviewed 19 existing industry governance frameworks and academic works to examine if the identified issues can be addressed by them. However, there are common missing considerations of how to clarify the rights of data owner or subject and how to achieve visibility and traceability in the use of data. Through the studies, we confirmed the need for a data governance framework for PEs to support the organizations' business success.

3 Methodology

This study was conducted through three steps (Fig. 1). The first step was carried out to understand overall PE environment including who participates in a platform, how data is used in the platforms and what data characteristics are identified (①). We analyzed five PEs: Facebook (social network), YouTube (content portal), EBay and Uber (exchange platform), and Data.gov.au (public platform). The survey on the platforms were conducted by examining their policies such as data polices, privacy policies or cookies polices. We carried out a literature review to complement and confirm the result of the analysis. The second step was to identify decision domains and governance principles (②). In the previous study [10], we identified seven decision domains through reviewing literature, industry governance frameworks and the state of practice of four platforms. In this study, we refined them and identified focal principles to support the decision domains. Based on the results of the two steps, we defined due processes for the implementation of data governance. To confirm the processes, we analyzed data breach cases (AOL and Facebook) and reviewed the relevant literature. In the last step, we compared 19 existing frameworks and academic works with our framework (③). We included some IT/information frameworks for this comparison because they generally contain data governance. We used the identified decision domains, principles and due processes as the comparison factors.

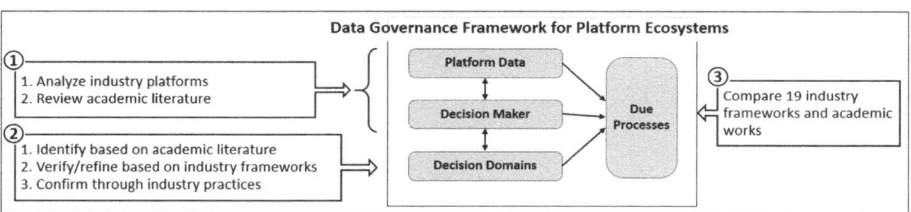

Fig. 1. The research methodology for a data governance framework for PEs

4   Data Governance Framework

4.1   The Conceptual Model

We start by presenting the general concept of data governance. There is a broad consensus among researchers that data governance must find answers to three questions: what decisions need to be made, which roles should be involved in the decision-making process and how the roles are involved in the process [11, 12]. Governance processes play a focal role to orchestrate and explain how the elements work together.

We develop a data governance framework for PEs based on the concept. It includes three core elements (data, decision makers, and decision domains) and due processes to support the elements. All the elements are discussed with a set of questions such as how data is collected and used in a PE, what decisions should be implemented about the data, and who can play the roles (Figure 2). The framework enables the governance body of a PE to find the answers of the questions by showing how platform data should be managed in data governance through due processes.

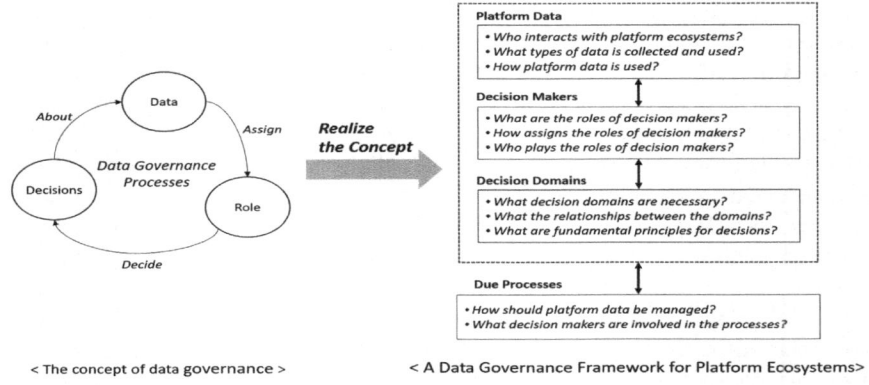

Fig. 2. A data governance framework for PEs from general concept

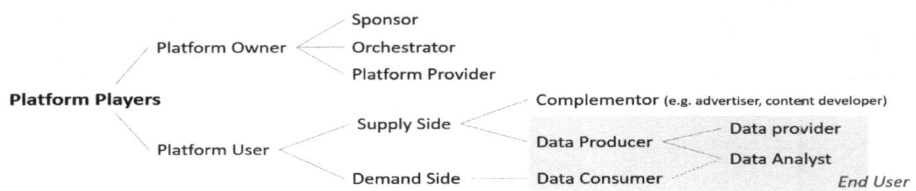

Fig. 3. Generic platform players in PEs

4.2   Platform Data

Ecosystem refers to a complex set of relationships among the elements of a given area. In a PE, data interacts with multiple participating groups. We begin with introducing platform player who interacts with a platform (Fig. 3), and then describe how platform data is collected and used, and discuss the current issues.

In general, platform player is divided into platform owner and platform user.

Platform owner consists of three roles: sponsor, orchestrator and provider. Platform sponsor owns a platform, and facilitates the co-creation of value from third-parties or establish an exchange platform he can benefit from [9]. The role of orchestrator is to organize a platform and the involved parties and processes. It is in charge of sharing standards, developing the industry vision or maintaining the integrity of a platform [4]. Platform provider is intermediary who delivers a platform. It generally includes the roles of data manager such as data collector, steward and custodian.

Platform user comprises supply and demand side. Data supply can be capable by complementor or data producer. Complementor contributes to a PE as an external party not directly related to the platform owner [9]. It offers a complementary content to the core component of the platform. Data producer consists of data provider who directly contributes data and data analyst who uses and provides data through data analytics jobs. On the demand side, data consumer refers to end user that uses platform data. Data analyst can be both data consumer and data provider if providing the outcome of the analytics jobs to the platform again. End user here is a person who accesses the platform to consume a service available on the platform [9, 15].

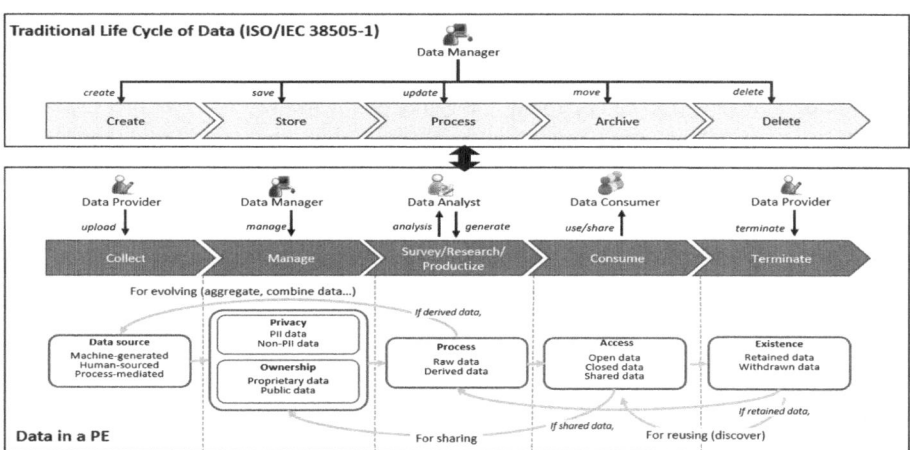

Fig. 4. Life cycle, characteristics and interactions with actors of data

The described roles can be changed over time or depending on platform strategy.

In traditional data governance, the life cycle of data is aligned with accountability for data management within an enterprise [16]. There is a simple interaction with a data manager. Meanwhile, the life cycle of platform data is based on the processes of data sharing. Platform data interacts with various platform players, and the flow map is characterized along the life cycle (Fig. 4).

Platform data is collected through providers' contribution such as uploading or generating new data. Majority data is from platform users as they upload their content such as video or image or provide user information (human-sourced data) [17]. While a user uses platform services, the platform systems may leave some data such as logs, search keywords, location (machine-generated data). This type of data is generally referred to service use information. Data is also collected through system processes such as transactions, reference tables, and relationships, as well as the metadata setting the context (process-mediated data).

The gathered data has to be examined whether it is Personal Identifiable Information (PII) data or non-PII data, and it is proprietary data or public data. Based on that, the management processes of data and the involved roles should be differed. PII is defined by Australian Acts as "Information or an opinion about an individual, or an individual who is reasonably identifiable, whether the information or opinion is true or not; and whether the information or opinion is recorded in a material form or not" [18]. However, PII and non-PII are not immutable [19]. If there are only single instances of users, it is easier to be identified by combination of the characteristics revealed in the datasets. To reduce the risk, continuous review process is necessary in the data management process. Due processes allow only expected activities of actors, and support identifying audit trails, offering interactive modeling and supporting user objections [8]. Proprietary data is claimed ownership by a specific entity or company. The owner of data should have certain decision rights and obligation about the data. In contrast, public data (e.g. crime data) is available for the public to collect or look at. As mentioned, the ownership of data is not clearly defined in platform policies. Our survey on four industry platforms [10] showed that Facebook, YouTube and Uber define data ownership of user content, but non-user content is rarely addressed. EBay documents overlook data ownership. To cope with the issue, platform owners have to consider the regulatory environment of a platform and determine an appropriate governance configuration prior to the use of data. It can reduce the risk of data misuse or abuse and protect the rights and privacy of the owner or subject of data.

The stored data in the platform systems can be internally used for their business to get useful information such as trends, statistics, significant keywords, or personal interests of users. The data can be evolved by aggregating or combining raw data, and then generates new data as derived data. 11 common use cases of platform data are identified through survey on industry platforms: provide, improve and develop (test) services, communicate with platform users, or show and measure ads and services. The cases, however, are not detailed documented in the policies of platforms. The data used for each use case is not precisely mentioned. It can result in data misuse. The issue is claimed by a number of researchers [8]. Another risk can be found when the stored data is used outside of a platform for survey or research by external partners. The exporting data should be reviewed by an appropriate policies if it can violate ownership or privacy rules or includes PII data. AOL and Facebook data breach (2006 and 2008) are reported as representative cases which the processes were ignored [1, 2].

General platform users can use platform data if data is set to "open" to everyone (open data) or a specific group or person (shared data). If a user changes the mode of his/her data into "private" (closed data), no one can use/access the data. Facebook documents that when a user post on Facebook, the user can select the audience for the post, such as a customized group of individuals, all of his or her friends, or members of a group. The platform mentions that open data is available to anyone on or off

Facebook services and can be seen or accessed through any online and offline media including search engines and TV. Open or shared data can encounter more risks of data abuse, misuse or privacy violation if there is no strong regulation or any complemental consideration. How to use, share or sell data without losing control is a critical issue of PEs [7]. Accordingly, data use cases should be explicitly defined in data governance as monitoring the use of data have to be implemented based on the use cases. A data supply chain also should be recorded to trace the derivation history of the open or shared data transparently. Such governance mechanisms should be fairly applied, and the processes and performance must be transparently shown to every participating group (in particular, to data owner and subject). Yet, the requirements are poorly implemented in industry PEs [10]. There are also claimed issues of an invisible data supply chain by researchers [8].

There is a broad consensus that data provider must have the privilege to stop sharing his data at any time. PEs provide several ways to change the mode of data sharing. A platform user can change the mode of data into private to stop sharing. The other way is to delete the data. Alternatively, the user can delete or deactivate the account. In theory, it looks as if data owners can perfectly control their data. However, in some cases, data owners lose control over their data. For instance, Facebook policies note that "information that others have shared about you is not part of your account and will not be deleted when you delete your account". That is, the shared data will be retained in the platform in the state that the owner is out of control, and continuously used/discovered by others. This issue has been discussed by researchers [20, 21]. Platform users' need for data transparency is increasing to access information which they are involved in. In addition, a certain method should be available to them for appropriate notice, consent and security.

4.3   Decision Makers

A typical data governance structure is organized within an enterprise. There is a lack of concerns about external users. In PE business context, platform users provide data, and it results in adding value to the platforms. Accordingly, they have a critical role in data governance of PEs [4]. In this sense, how to share the roles of decision making about data assets with platform users becomes an important issue [22]. We identify the key roles of decision makers for PEs including platform users: data committee/council, data manager, data owner and data subject.

Data committee/council is one of the role which is responsible for clarifying the role of data in PEs [11]. It makes decisions about the purpose of data use, desirable behaviors, and the appropriate governance mechanisms of a PE aligning the business goals. The role is generally taken by platform orchestrator.

Data manager here refers to the role of internal data management in platform companies including data collector, data steward and data custodian [11, 23]. They are responsible for the implementation of data management tasks and the conformance to governance rules in platforms. Data governance design can be categorized into centralized and decentralized [12]. While centralized governance means a platform owner takes all the control, decentralized governance shares it with platform users. Therefore, some parts of the role can be implemented by the users in decentralized governance.

Data owner is an individual (or a company) who owns data by contributing it to the platforms. Data owner has ownership rights which refer to the questions of who is allowed to use data and who has decision rights [24]. Accountability of a data owner is noted as a form of verifiability in some literature. The term verifiability represents a sort of responsibility of the one who can verify data and confirm the veracity of the data before using or sharing the data [20]. Data owners should have data transparency and auditability, and access control power [21]. That is, every user has complete transparency over what data is being collected about her and how the data is used.

Data subject means a person who is the subject of personal data. If data is about a specific person, then that person can be a data subject. Data owner can be a data subject and vice versa if he/she uploads/generates data about him/her. There is an example to explain the difference between data owner and data subject in a simple way. A medical record of a patient is generated by a doctor/hospital. The owner of the record is the doctor/hospital that generated the record. The patient is a data subject because the medical record is about him, but he cannot own the record. Like a data owner, a data subject should have rights to access the data which he is involved in and a method available to him to hold data governance mechanisms accountable for appropriate notice, consent and security.

The described roles can be taken by various platform players depending on the platform strategies (Table 1). In decentralized data governance, platform users can monitor or audit the use of data or data integrity based on enabling technologies [25].

Table 1. The roles of decision makers and platform players

Role of decision makers		Platform player							
		Platform owner			Platform user				
					Supply side			Demand side	
Role	Description	Sponsor	Orchestrator	Provider	Complementor	Data provider (Data producer)	Data analyst	Data consumer (General user)	
Data committee	Clarify the role of data for platform business	√	√						
Data manager	Collect, create and manage data based on the defined processes				√				
	Monitor and audit the activities taken place based on the policies				√	√ (If a platform shares control power with the platform users)			
Data owner	Analysis and generate new data by using existing data						√		
	Upload content/personal information like video, image or user profile				√	√			
	Invest resources and IT services to maintain non-IP content	√		√					
Data subject	Upload (post) data/information about himself/herself				√	√			
	Upload (post) data/information about others				√				

4.4 Decisions Domains

Decision domains refer to data governance areas which should be controlled to achieve the business goals of a PE. In our previous study [10], we identified seven data governance factors which can be used as decision domains for PEs (Table 2).

The decision domains are interconnected with each other. When platform data is used, there should be clear definition of the roles about the data such as who has accessibility or accountability, and who should be informed or consulted. A data ownership and access definition are thus regarded as a major concept when designing the business process of a PE [9, 22]. The definition should include user content and non-user content together to protect all the data and owners or subjects' rights against unauthorized use. To support this, a data ownership decision model should be developed by considering relevant regulations, laws or court cases [11]. For example, creativity (creative data: videos/non-creative data: factual data), originality (original data/derived data), investment (data managed by a platform owner or not) and source (from outside or created inside of a platform) of data can be the aspects of the model. They are derived from the review of regulatory environment such as Berne Convention and its derivatives, European Court of Justice (ECJ) in 2004 (William Hill case [26]) and the policy of platforms. Looking at the regulatory environment of a platform also supports accurately identifying and rewarding the contributors of a platform as it clarifies who adds value to the platform. The main role of data committee is to build policies for a platform based on the review on regulatory environment. The policies have to include all the considerations of how to use data, what data can be open (or not), how to share data or how to terminate data sharing.

Table 2. Data governance decision domains for PEs

Decision domain	Definition
Data ownership/access	Definition of who owns, uses and accesses platform data
Regulatory environment	Regulations, laws or court cases that could affect the use of data
Contribution measurement	Mechanisms to measure contribution against value creation to a PE
Data use case	The purpose of the collected data by a PE (how to use data)
Conformance	An audit for compliance based on strict processes and rules
Monitoring	Mechanisms to monitor the use of data (all activities related to data)
Data provenance	Means to trace the derivation history of the data transparently

When data is collected and used by a platform (platform users), if there is an only single owner, contribution measurement is simple. Meanwhile, using derived data (aggregated or transformed data) can lead to measurement issues because the data may contain a complicated ownership structure. Data provenance management can help this issue. It allows a platform to identify all the associated stakeholders and explicitly measure the contribution of each owner of the data by preserving all the record of the use of data. It also supports high visibility of the use of data [11, 27].

As stated, the purpose of data uses and the relevant data are not clearly defined in the policies of PEs. The documentation is not enough to understand how the collected data is used. All the collected data should be categorized and has a clear and limited purpose of the use of data. It enables a platform to detect and prevent unexpected use of data in a data supply chain [16].

Monitoring and conformance mechanisms facilitate visible/reliable data use. There are many data breaches caused by an invisible supply chain and unclear due processes

[1, 2]. To increase transparency of a platform and thus gain more trust from platform users, a platform owner can share control power and decision rights with the users through decentralized data governance [25].

To support right decisions, data governance should be implemented based on key principles which present sets of applicable guidelines and considerations. Through a literature review and survey on industry platforms, the following four principles are identified, which have been regarded as fundamental considerations.

Table 3. The influence of the data governance principles on the decision domains

Decision domain	Principle 1 Align with platform governance concepts	Principle 2 Meet the needs of all participating groups	Principle 3 Address all types of data	Principle 4 Consider platform context
Data ownership, access definition	Define clear roles and responsibilities	Consider all data contributors' needs and rights	Clarify ownership and access rights to all types of data	Apply different levels of governance control based on the context of a platform * Highly regulated environment, high quality of data strategy, closed platform strategy or authorized-based governance configuration -> use strict data ownership, access control, audit and monitoring by a centralized (internal) structure * In the opposite case -> share the control power with platform users and use trust-based control
Regulatory environment	Identify what regulations should conform to control	Develop a decision model for explicit data ownership	Consider extensive regulations for non-user content	
Contribution measurement	Consider a revenue sharing concept	Identify different types of contribution of participants	Measure every data contribution based on regulations	
Data use case	Build trust through a visible data supply chain	Consider how to use data without losing control	Provide a detailed data category and use cases	
Conformance	Conform governance rules through a regular audit	Involve various participating groups	Audit every data use case and its processes	
Monitoring	Control an unauthorized data use	Provide possible opportunities to all stakeholders	Make a visible supply chain for all data use activities	
Data provenance	Support efficient, effective control and clear roles	Enable data owners and subjects to trace the history of the use of data	Record all the use of data including sharing, analyzing and transforming	

Principle 1. Align with platform governance concepts and business goals. Data governance goals can be identified and determined by looking at what to maximize the value of data and a PE. The goals, thus, should align the business goals and higher-level governance like platform governance [11, 28]. The characteristics of a platform also can be considered when confirming data governance goals. If platform open strategy leans to close, the data governance should be toward the focus on strict due processes and input/output control mechanisms [25].

Principle 2. Meet the needs of all participating groups. A PE faces the complicated relationships between multiple parties. Trust between platform owners and the parties is regarded as a prerequisite factor to win business [9, 14]. It can be built by starting with a good understanding of what governance practices are applicable and how they work, and share value (management strategies of a platform). Accordingly, data governance should be designed and implemented from all the perspectives of parties.

Principle 3. Address all types of data. Data governance should be able to control all types of data in platforms. As mentioned, platform data is collected from various source. Yet, PEs are mainly focused on user content [10]. The other types of data are often ignored and thus do not addressed in data governance processes. It leads to unclear data ownership or access rights of data owner or subject.

Principle 4. Consider platform context; one size does not fit all. Platforms have to consider different business strategies and goals, and consider different levels of market regulation. According to contingency theory, such different contingencies affect data governance [12]. In the previous study, we examined the influence of specific platform contingencies on the characteristics of a platform and a data governance design [25]. This principle gives the idea that data governance can be flexible based on the context of a platform and tailored for practical implementation.

The principles affect the decision domains in a certain way. They help a platform to focus on the key considerations and ultimately enable a platform to win business. Table 3 shows how the principles are applied to the decisions domains.

4.5   Due Processes in the Use of Data

Due process is regarded as a pivotal control mechanism to mitigate a risk of data abuse or misuse as it forces desirable behavior of participants [8].

In this section, we suggest comprehensive due processes in the use of data to show how platform data should be managed in data governance. All the considerations and discussions in the previous sections are deliberated in the processes. The processes also illustrate how the roles described previously are involved and how the decision domains are implemented in data governance. This section sequentially demonstrates the processes by following the lifecycle of platform data (Fig. 4).

Data Collection Process (Fig. 5). Data collection is implemented by defining data categories and data use cases based on the principles and policies of a platform. When data is collected by a platform, it should be classified by the defined data category. The use of the data thus can be limited by the predefined and linked use cases to prevent illegal use of data (❶). The data is also characterized based on the source. It can be used

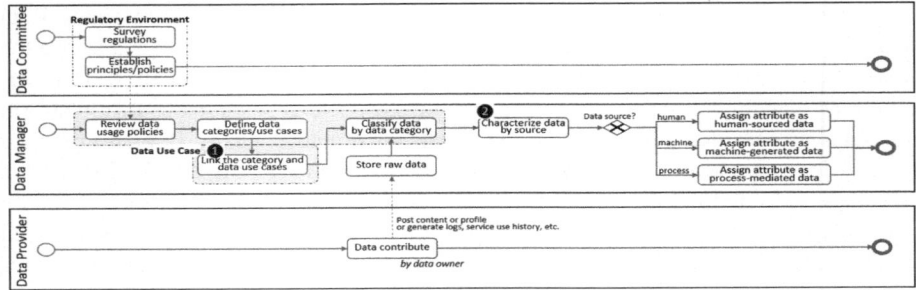

Fig. 5. Data collection process

for data ownership recognition (❷). In general, human-sourced data is regarded as Intellectual Property (IP) content, and it belongs to the provider.

Data Management Process (Fig. 6). The main focus of this process is on the rights protection of data owner and subject. Data committee has to establish clear policies in terms of data ownership, access right and privacy based on the relevant regulations or laws. For privacy protection, every stored data should be tested (PII or non-PII data). According to the result, the data needs to be dealt by the different levels of processes and policies (❶). A model for ownership definition is developed by following the defined policies of a platform (❷). It is used when identifying the owner of data to measure contribution and assign explicit roles and responsibilities. The information should be sent to the owner and subject of the data (❸). Data provenance can be initialized for recording historical information of the data use (❹).

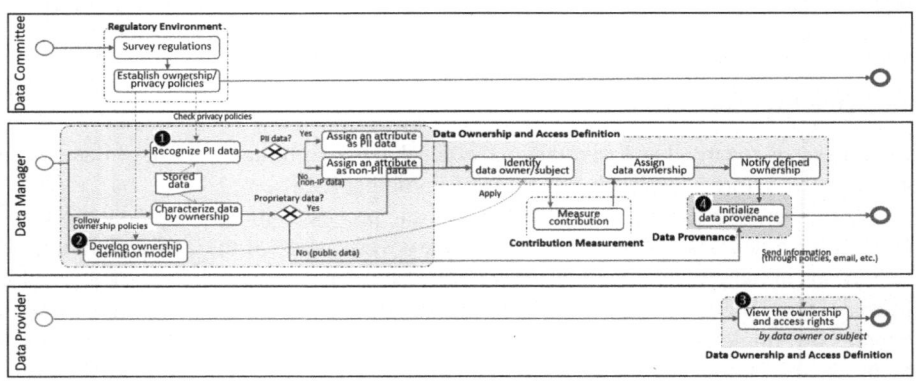

Fig. 6. Data management process

Survey, Research and Productization Process (Fig. 7). Platform data can be used for improving the services of a platform company. In addition, it can be required for external use such as research purpose. In those cases, first of all, every access should be confirmed if it is legal and the purpose of the use meets the predefined use cases of the

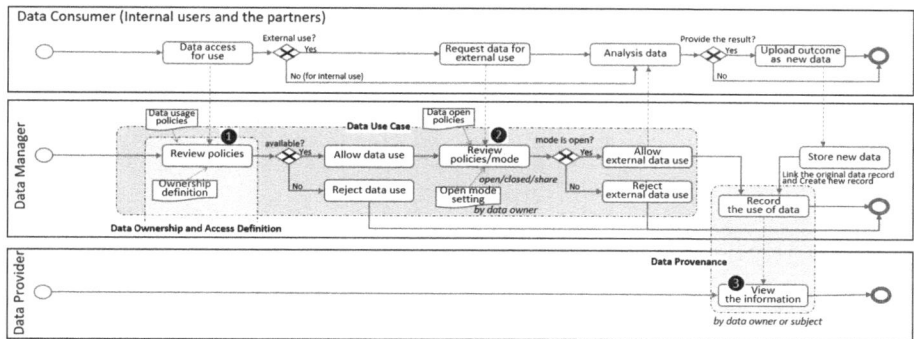

Fig. 7. Data survey, research and productization process.

data (❶). Secondly, if the data is taken out and possibly disseminated for secondary use, the openness of the data and platform policies must be checked (❷). Facebook data breach happened as the company overlooked the process [2]. A group of personal information was exported for a research project without a review process, and quickly diffused for secondary use. It resulted in revealing the data to public without consent of the data owners and subjects. Lastly, the data owners and subjects should transparently know all the information of the use of their data to support user objections (❸).

Data Consumption Process (Fig. 8). The open or shared data in a platform can be discovered and used by other users (❶). Like the previous process (Fig. 7), all the processes should follow the relevant policies and be reviewed. This process pays more attention to high participation of platform users and transparency of a platform. When a platform company shares control with platform users (in decentralized data governance), platform users can actively participate in auditing or monitoring data and data use processes (❷❸). It is made possible by enabling technologies such as blockchain which is regarded as one of the most innovative and revolutionary governance forms [29]. This process enables an organization to reduce cost and effort, and gain more trust between a platform owner and the platform users [25].

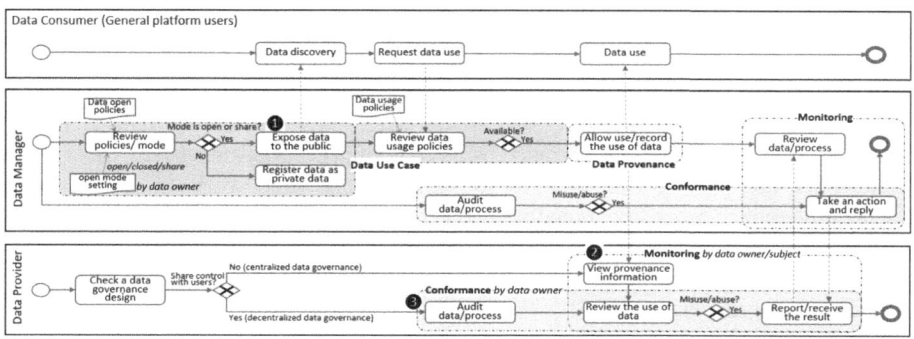

Fig. 8. Data consumption process

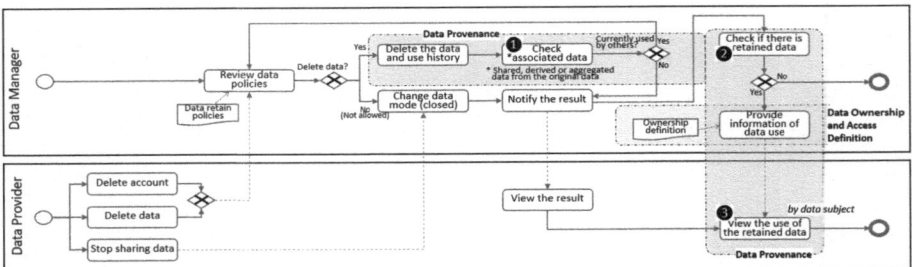

Fig. 9. Data sharing termination process

Data Termination Process (Fig. 9). When a user deletes his account or content, the content may be deleted from the platform systems. Depending on the data retention policies of a platform, the deleted data can be retained for a certain period of time, but it is ultimately deleted. However, the shared or derived data of the user can be retained and out of control of the owner (❶). Accordingly, if data is retained based on the policies of a platform, even though the owner lost the ownership, the rights of the data subject to the data should be protected and respected (❷). In this sense, the information of the use of the data must be accessible by the subject (❸).

5 Comparison

We compare 19 governance frameworks and academic works with our framework which were used in the previous study for the state of the art (Table 4). ISO/IEC 38500 is replaced with 38505-1 as it has been released as a data governance standard.

We use the principles (P), decisions (D) and due processes (DP) presented in this study as comparison factors to evaluate the comprehensive concerns and impact of the compared frameworks and studies (P and DP are added to the previous analysis). "Sufficiency" is used to examine if the factors are dealt in the frameworks [10]. A result is determined as "not covered (x)", "partially covered (○)", or "covered (●)".

The result reveals three main findings. Firstly, any framework or study that covers all the considerations addressed in our framework have not been found. Most of IT/data governance frameworks focus on general roles and responsibility (D1) or role definition and control by governance bodies (D1, 5 and 6). Secondly, platform studies pay more attention to the concept of PE, and platform control mechanisms (D5 and 6). As the studies are still at a relatively embryonic stage, how to manage data is largely neglected. Lastly, while the importance of governing process is stressed by most of the frameworks, due process has not been suggested in any framework. COBIT 5.0 documents governance processes, practices and activities but how organizations implement the processes and what roles should be involved in the processes are not described. It may lead that organizations have difficulties in newly applying or improving data governance in practice.

We confirm that there are significant gaps between the compared frameworks and our suggestion which should be filled. It shows the need for our framework again.

Table 4. The result of comparison of governance frameworks

Category	IT Governance				Information/Data Governance					Governance for Platform Ecosystems							Our Framework
	COBIT 5.0	Weill & Woodham (2005)	Weill & Ross (2004,2005)	ISO/IEC 38505-1	DGI Framework	Informatica Framework	IBM Information Governance	Khatri & Brown (2010)	Weber et al. / Otto & Weber (2009, 2015)	Evans (2012)	Glazaruch & Heathnlsson (2010, 2013)	Hagiu (2014)	Manner et al. (2012)	Manner et al. (2013)	Tiwana et al. (2010)	Tiwana (2013)	
P1: Align platform governance	●	X	○	○	X	X	●	●	X	X	X	X	X	○	X	X	●
P2: Meet the needs of all participants	●	X	○	○	X	○	X	○	X	X	X	X	X	X	X	○	●
P3: Address all types of data	X	X	X	X	X	○	X	X	X	X	X	X	X	X	X	X	●
P4: Consider platform context	X	○	X	X	○	X	X	X	●	X	X	X	○	○	X	X	●
D1: Data ownership and access definition	○	○	○	○	●	●	○	○	○	X	X	X	X	X	○	○	●
D2: Regulatory environment	○	○	○	●	X	X	○	○	X	X	X	X	○	X	X	X	●
D3: Contribution measurement	X	X	X	X	X	X	X	X	X	X	X	X	X	X	X	X	●
D4: Data use case	X	○	○	○	○	X	X	○	●	X	X	X	X	X	X	X	●
D5: Conformance	●	X	X	●	X	●	○	○	X	○	○	○	X	○	○	○	●
D6: Monitoring	○	○	X	●	○	●	○	○	X	○	X	X	○	○	○	○	●
D7: Data provenance	X	X	X	X	X	●	○	○	X	X	X	X	X	X	X	X	●
DP: Collection, management, analysis, consumption and termination process	○	X	○	X	○	○	X	X	X	X	X	X	○	X	X	X	●

6 Conclusion and Future Work

Many organizations today adopt or consider PE for their business innovations. The concept of PE supports sustainable growth through network effects where there is multiple groups' data contribution. However, lack of organizational capability to orchestrate complicated context, processes and relationships occurred among the parties will lead to market failure. Traditional data governance focuses on in-house control of data, and prior research on platform a governance is still in its infancy.

In this study, we proposed a new data governance framework which supports an organization to mitigate business risks from the complexity of a platform and add value to the organization. We surveyed industry platforms and reviewed governance frameworks and literature. This study delivered the idea on how data should be managed when an organization adopts the concept of platform ecosystem. In particular, through the due processes, we demonstrated how organizations can implement data governance and orchestrate all the considerations of platform ecosystem. We compared the framework with 19 existing industry governance frameworks and academic works. The comparison showed that there is no existing framework or study which covers all the aspects of our suggestion in the framework.

In the next step, we will provide the use cases of the framework to assist an organization to implement data governance in practice. We will identify use case scenarios and the associated governance questions for decision-making which are critical but cannot be answered by current governance frameworks. To this end, we will perform an extensive literature review and survey industry needs.

References

1. Ives, B., Krotov, V.: Anything you search can be used against you in a court of law: data mining in search archives. Commun. Assoc. Inf. Syst. **18**(1), 29 (2006)
2. Zimmer, M.: "But the data is already public": on the ethics of research in Facebook. Ethics Inf. Technol. **12**(4), 313–325 (2010)
3. Parker, G., Van Alstyne, M.W.: Platform strategy. In: The Palgrave Encyclopedia of Strategic Management (2014)
4. Smedlund, A., Faghankhani, H.: Platform orchestration for efficiency, development, and innovation. In: 2015 48th Hawaii International Conference on System Sciences, pp. 1380–1388 (2015)
5. Kaisler, S., Armour, F., Espinosa, J.A., Money, W.: Big data: issues and challenges moving forward. In: 2013 46th Hawaii International Conference on System Sciences, pp. 995–1004 (2013)
6. Kaisler, S., Money, W.H., Cohen, S.J.: A decision framework for cloud computing. In: 2012 45th Hawaii International Conference on System Sciences, pp. 1553–1562 (2012)
7. Jagadish, H.V., Gehrke, J., Labrinidis, A., Papakonstantinou, Y., Patel, J.M., Ramakrishnan, R., Shahabi, C.: Big data and its technical challenges. Commun. ACM **57**(7), 86–94 (2014)
8. Martin, K.E.: Ethical issues in the Big Data industry. MIS Q. Executive **14**, 2 (2015)
9. Schreieck, M., Wiesche, M., Krcmar, H.: Design and governance of PEs–Key concepts and issues for future research. In: 24th ECIS 2016 (2016)
10. Lee, S.U., Zhu, L., Jeffery, R.: Data governance for PEs: critical factors and state of the practice. In: 21st PACIS 2017, Malaysia (2017)
11. Khatri, V., Brown, C.V.: Designing data governance. Commun. ACM **53**(1), 148–152 (2010)
12. Weber, K., Otto, B., Österle, H.: One size does not fit all—a contingency approach to data governance. J. Data Inf. Qual. **1**(1), 1–27 (2009)
13. Evans, D.S.: Governing Bad Behavior by Users of Multi-Sided Platforms, SSRN Scholarly Paper No. ID 1950474. Social Science Research Network, Rochester, NY (2012)
14. Hein, A., Schreieck, M., Wiesche, M., Krcmar, H.: Multiple-case analysis on governance mechanisms of multi-sided platforms. In: Multikonferenz Wirtschaftsinformatik (2016)
15. Tiwana, A., Konsynski, B., Bush, A.A.: Platform evolution: coevolution of platform architecture, governance, and environmental dynamics. Inf. Syst. Res. **21**(4), 675–687 (2010)
16. ISO. https://www.iso.org/standard/56639.html. Accessed 27 Sept 2017
17. Firmani, D., Mecella, M., Scannapieco, M., Batini, C.: On the meaningfulness of 'Big Data Quality' (Invited Paper). Data Sci. Eng. **1**(1), 6–20 (2016)
18. Australian Government. https://www.legislation.gov.au/Details/C2018C00034. Accessed 3 Oct 2017
19. Schwartz, P.M., Solove, D.J.: The PII problem: privacy and a new concept of personally identifiable information. NYUL Rev. **86**, 1814 (2011)
20. Al-Khouri, A.M.: Data ownership: who owns "my data". Int. J. Manag. Inf. Technol. **2**(1), 1–8 (2012)
21. Zyskind, G., Nathan, O.: Decentralizing privacy: Using blockchain to protect personal data. In: Security and Privacy Workshops (SPW). IEEE (2015)
22. Tiwana, A.: PEs: Aligning Architecture, Governance, and Strategy, Newnes (2013)
23. Cheong, L. K., Chang, V.: The need for data governance: a case study. In: ACIS 2007 Proceedings, vol. 100 (2007)

24. Eckartz, S.M., Hofman, W.J., Van Veenstra, A.F.: A decision model for data sharing. In: Janssen, M., Scholl, H.J., Wimmer, M.A., Bannister, F. (eds.) EGOV 2014. LNCS, vol. 8653, pp. 253–264. Springer, Heidelberg (2014). https://doi.org/10.1007/978-3-662-44426-9_21
25. Lee, S.U., Zhu, L., Jeffery, R.: Designing data governance in PEs. In: 2018 the 51st HICSS, Hawaii, (2018)
26. Harison, E.: Who owns enterprise information? Data ownership rights in Europe and the US. Inf. Manag. **47**(2), 102–108 (2010)
27. Informatica, https://www.informatica.com/content/dam/informatica-com/global/amer/us/collateral/white-paper/holistic-data-governance-framework_white-paper_2297.pdf. Accessed 14 Sept 2016
28. Weill, P., Ross, J.W.: IT Governance on One Page, SSRN Scholarly Paper No. ID 664612, Social Science Research Network, Rochester, NY (2004)
29. Parker, G., Van Alstyne, M.W., Choudary, S.P.: Platform Revolution: How Networked Markets are Transforming the Economy and How To Make Them Work for You (2016)

Designing a Process Mining-Enabled Decision Support System for Business Process Standardization in ERP Implementation Projects

Christian Fleig[✉], Dominik Augenstein, and Alexander Maedche

Institute of Information Systems and Marketing (IISM),
Karlsruhe Institute of Technology (KIT), Karlsruhe, Germany
christian.fleig@kit.edu

Abstract. Process standardization allows to optimize ERP systems and is a necessary step prior to ERP implementation projects. Traditional approaches to standardizing business processes are based on manually created "de-jure" process models, which are distorted, error-prone, simplistic, and often deviating from process reality. Theoretically embedded in the organizational contingency theory as kernel theory, this paper employs a design science approach to design a process mining-enabled decision support system (DSS) which combines bottom-up process mining models with manually added top-down standardization information to recommend a suitable standard process specification from a repository. Extended process models of the as-is process are matched against a repository of best-practice standard process model using an attribute-based process similarity matching algorithm. Thus, the DSS aims to reduce the overall costs of process standardization, to optimize the degree of fit between the organization and the implemented processes, and to minimize the degree of organizational change required in standardization and ERP implementation projects. This paper implements a working prototype instantiation in the open-source process analytics platform Apromore based on a real-life event log and standardization attributes for the Purchase-to-Pay and Order-to-Cash processes from three SAP R/3 ERP systems at the industry partner.

Keywords: Process mining · Decision support systems
Process standardization · Enterprise resource planning
ERP implementation projects

1 Introduction

Rapidly evolving competitive environments and emerging business opportunities require the transformation of business processes in the organization in response to new conditions to remain competitive [1]. However, the transformation of a business process from a current design to a target process design requires organizations to precisely understand the real-world execution of the as-is process to make solid transformation decisions (e.g., [2]). Organizations frequently do not to meet these prerequisites for business process transformation, and possess only limited insights and a narrow

© Springer Nature Switzerland AG 2018
M. Weske et al. (Eds.): BPM Forum 2018, LNBIP 329, pp. 228–244, 2018.
https://doi.org/10.1007/978-3-319-98651-7_14

understanding of existing process execution paths [3]. Traditional top-down approaches to business process transformation rely on "de-jure" process analyses instead of bottom-up "de-facto" data-driven approaches. These "de-jure" approaches suffer from a number of insufficiencies as they are based on handmade process models which are often biased compared to process reality [4]. "De-jure" process documentations usually only contain idealistic process executions such as the to-be process, while most process variants and deviations from the ideal target specification are ignored [5]. In addition to content-related insufficiencies, top-down process modeling itself is a time- and resource-consuming task [6]. Further, "de jure" process models are error-prone due to their manual creation. In sum, van der Aalst finds that the currently prevailing approaches of process modelling are "disconnected" from process realities [7], which implies that human-centered top-down approaches provide only an insufficient base for decision-making in process transformation.

A chance to overcome these weaknesses of decision-making in process transformation is to utilize the increasing availability of process data from numerous information sources in organizations [8]. For example, information systems store process events in large event log tables [9] which provides the possibility to improve decision-making by data-driven approaches to process analytics such as process mining [5]. For example, process mining delivers descriptive and positive "de-facto" process analyses based on bottom-up data [5]. Hence, "de-facto" process analyses provide a valuable complement to decision-making in process transformation.

As a particular field of process transformation, business process standardization has experienced a high degree of scholastic attention [10], and has been recognized as a critical step prior to the implementation of new enterprise resource planning (ERP) systems (e.g., [11]). However, ERP systems such as SAP or Oracle provide numerous alternatives of possible standard processes by software vendors. In workshops performed at our industry partner in the context of a large-scale business process standardization and SAP S/4 HANA implementation project, we discovered that organizations are frequently challenged by the selection of the most appropriate standard process design. Thus, organizations might significantly benefit from a decision support systems (DSS) in the selection of suitable standard processes which overcomes the outlined weaknesses, and which considers the very specific process requirements of the individual organization. The research question of this paper therefore becomes:

How to design a process mining-enabled decision support system to support organizations in the standardization of business processes?

Besides a practical need for data-driven standardization decisions, an important research gap refers to the absence of contributions on the "post-mining" phase, with only few contributions exploring the question of how to turn the insights gained by process mining into actual process transformation decisions. This paper employs a design science research (DSR) approach to motivate, conceptualize, develop, and to evaluate the DSS artifact.

The rest of this paper is organized as follows. Section 2 briefly introduces conceptual foundations. Section 3 describes the design science research (DSR) methodology to systematically derive and implement a working DSS in the Apromore process analytics platform [12]. Section 4 derives meta-requirements (MRs) for the DSS which serve as developmental guidelines. Section 5 further concretizes the MRs in design principles (DPs) and design decisions (DDs) to develop the technical blueprint conceptualization. Section 6 describes the implementation in Apromore based on data from three real-world SAP R/3 ERP systems for the purchase-to-pay ("Purchasing") and order-to-cash ("Sales") processes from a manufacturing corporation. Section 7 presents results from three evaluations of different aspects of the DSS. Section 8 concludes and presents limitations and avenues for future research.

2   Related Work

This section lays the literature foundations for the design requirements to the process mining-enabled DSS. To achieve the intended purpose, the DSS requires theoretical embedding in literature on process standardization, process mining, and similarity-based process matching. This section introduces the kernel theory from organization science as theoretical embedding and presents related work for the derivation of meta-requirements and design principles and decisions in Sects. 4 and 5.

This research is theoretically motivated by organizational contingency theory by Donaldson [13] and Sousa and Voss [14], which requires organizations to adapt structures to maintain a fit between changing contextual factors and environmental variables to retain performance. With particular regard to the field of BPM, as business processes are highly context-dependent (e.g., [15]) and business processes are systems which interact with the environment [16]. Extant research such as the contribution by vom Brocke et al. finds the effect of process management to be contingent upon contextual factors including organizational factors, process characteristics, and goals [17]. Thus, contingency theory requires a fit between business processes and environments [18], and to adapt business processes in response to any change in environmental variables. In addition to generic contingency theory, the organizational information processing theory by Galbraith [19] considers organizations as information-processing units which collect and process information and thus need to fit variables inside and outside the organization [20]. Therefore, the DSS designed in this research further incorporates contextual process factors such as standardization attributes to yield standardization support based on the contingencies of a particular organization.

We define a DSS as any system to address semi-structured or unstructured problems to support decision-making processes of users (e.g., [21]). Besides, Numerous contributions reveal a vital importance of alignment between the organization, business processes, and ERP systems [22]. Process standardization aims at a situation where the same activity in different organizational units is performed identically [23]. A standardized process "is constantly performed following the same steps in the same sequence" [24] and standardization can be achieved by the application of formalities, e.g. by creating guidelines or work instructions [24].

Contemporary information systems such as WfM, ERP, CRM, SCM, and B2B systems record business events in so-called event logs, which serve as foundations for process mining [3, 9]. For example, SAP logs all transactions, e.g., users filling out forms, changing documents etc. which significantly improves the ability to derive process transformation decisions by taking into account process variants and additional process information. To overcome the outlined weaknesses of top-down approaches to process standardization, this paper aims to utilize process to select standard processes using process mining as a source of bottom-up process information. Process mining aims to automatically discover business processes from transaction data [9, 25], and offers a spectrum of techniques to perform automatic process discovery, monitoring, and improvement activities using system data in event logs [4]. In particular, process mining retrieves process models, which graphically and analytically represent business processes [22] and depict the course of activities and their dependencies [26].

In addition to process mining, the DSS is required to perform a matching of the as-is process against best-practice standard processes to propose a suitable standard process for implementation. The application of similarity for process matching is motivated by the minimization of disruptiveness of the new future process design to the organization. ERP implementation projects impose a "technochange" situation on organizations as ERP projects simultaneously impact technological as well as organizational structures. Technochange situations require significant efforts in terms of IT project management and change management [22]. Hence, adequate to-be standard processes are characterized by a high degree of implementability. Implementability addresses limitations in the organizational adaptability, and thus requires a minimum of misfits to the organization [27]. Therefore, selecting business process designs X' which exhibit a high degree of similarity to the current as-is process in (X) reduces misfits of the selected and the status-quo business process. Misfits are the result of low similarity between the current business process and the future business process. The resulting transformation for a business process with a low level of similarity between X and X' requires large transformation efforts, which overhauls routines and modifies well-accustomed workflows. As a consequence, adverse technochange situations and risks might arise for the organization such as high costs, a reduction of organizational performance, or the avoidance of the information system when choosing a target process with a low degree of similarity to the as-is process.

3 Research Methodology

Design science develops artifacts to address important organizational problems [28]. Thus, this paper employs a design science research (DSR) approach to provide organizations with a "process mining-enabled DSS" in two design cycles in a "build-and-evaluate-loop" [28]. In addition to providing a software artifact, we aim to derive the design requirements as a theoretical contribution for the system to abstract from the concrete artifact. We conduct the DSR project within the context of a large-scale ERP implementation project, which comprises the replacement of the current SAP R/3 ERP by the future SAP S/4 HANA Business Suite. In 2017, the corporation consisted of several sub-companies operating globally with more than 8.200 employees and about 1.2bn Euro in turnover. The industry partner provided an event log for the purchase-to-pay ("Purchasing") and the order-to-cash ("Sales") process for three companies for period from 01/2016 to 07/2017. Therefore, this contribution therefore uses a real-life event log, and thus overcomes the weaknesses of many process mining contributions when relying on synthetic, simulated data. Project responsibility is allocated to a coordination team of senior decision-makers which serve as workshop participants to derive meta-requirements and design principles apart from literature. Each design cycle consists of a problem awareness, a suggestion, a development, and an evaluation phase as proposed by the seminal contribution by Hevner et al. [28]. In the problem awareness phase of cycle one, a structured literature on process mining and ERP implementation projects was conducted to validate the theoretical research gap and the need for decision support in business process standardization in the context of ERP implementation projects. An important gap in process mining research is characterized by the lack of research on the "post-mining phase", with almost no contributions investigating the question of how the findings from process mining can actually be used in the standardization of business processes to support ERP implementations. In the suggestion phase of the DSR project to address the research problem, meta-requirements and design principles are derived in four workshops. Participants include decision-makers from the different sub-companies, namely the chief information officer (CIO), the project leader (IT/ERP process expert), a leading operations manager (manager supply chain execution), a sales process expert (supervisor market research), a senior accountant (director controlling), an external IT and ERP consultant, and the PhD-student (first author of this paper in a passive form). Further, a literature review was conducted to enrich meta-requirements from practitioner workshops with theoretical foundations from the fields of process standardization, process mining, ERP implementations, and process matching techniques. In the development phase of cycle one, a DSS prototype was developed in Apromore [12]. As the entire system can hardly be evaluated in a single evaluation, the evaluation is split into different evaluations of individual system aspects. The first design cycle performs an evaluation of three aspects of the system. First, a technical evaluation is performed to demonstrate feasibility of the approach. Second, the system links process models with standardization attributes. These attribute-extended process models are evaluated in terms of the ability to increase process model comprehension of decision-makers. Third, semi-structured interviews are performed with decision-makers to determine system quality and usefulness in process standardization (e.g., [29]).

Design cycle two will consist of a refinement of meta-requirements and design principles to arrive at a final conceptualization of the DSS. The second design cycle will further concretize the design requirements to incorporate learnings from the evaluations performed in cycle one and to improve the Apromore artifact. In particular, following the demonstration of technical feasibility with real data in the Apromore instantiation, solid design science requires a further evaluation of the process model matching algorithm in future research. Findings in the evaluation from the previous design cycle will be implemented in the DSS instantiation to finalize the Apromore software artifact. Design cycle two will close with a second evaluation round in terms of whether managers would actually decide to adopt the DSS in projects.

4 Meta-requirements

Organizational process knowledge might either be stored in prescriptive "to-be" and top-down sources of information such as the implicit knowledge of process participants or be stored in descriptive "as-is" bottom-up sources such as information systems.

As each of these two types of process information has individual strengths and weaknesses, the DSS needs to be able to retrieve and combine process knowledge from different sources, and to combine these different types of process information before deriving decision support for business process standardization.

Thus, the DSS needs to incorporate bottom-up process information into decision-making to provide "as-is" process-specific standardization guidance. A potential source is bottom-up process information stored in information systems such as ERP systems. These sources include data generated by systems during process execution, such as event log tables within the ERP systems.

MR1: The DSS needs to incorporate de-facto bottom-up process information.

These data sources capture process executions "as-is". An exclusive reliance upon process mining in decision-making for business process standardization yields merely an incomplete picture of process realities. Process mining captures only information on process activities within the information system (e.g., [4]), and event logs merely contain a subset of all possible process facets [4, 5]. Therefore, insights gained from bottom-up sources might be incomplete due to shadow process steps which are not recorded in the system event log. The DSS needs to incorporate different types of quantitative and qualitative process information in addition to bottom-up models and additional process knowledge needs to be retrieved from top-down sources. In particular, top-down sources comprise intangible human process knowledge which cannot be retrieved bottom-up as these process elements are not executed within the information system. Examples include paper-based process steps, third-systems, inputs, outputs, off-system data, and participating user groups. We introduce MR2 accordingly:

MR2: The DSS needs to provide a user interface to retrieve additional top-down process information.

Organizational contingency theory by Donaldson [13] requires activities of business process management to consider the respective circumstances and contexts of business processes into decision-making. The work by Rosemann and Vessey [30] introduces the notion of context-dependent processes. Therefore, the DSS needs to incorporate relevant contextual process information and to capture information such as standardization goals, process type, and key process dimensions and characteristics to provide tailored decision support depending on the circumstances of the respective organization and the respective business process. We consequently introduce MR2a:

MR2a: The DSS needs to incorporate process context factors and process characteristics into decision-making.

Furthermore, most approaches in BPM usually incorporate strategic process goals [31] which are compatible with the overall organization strategy. These transformation goals serve as an input for the DSS to derive process transformation recommendations and to choose among alternative competing standards. We formulate MR2a:

MR2b: The DSS needs information concerning process standardization goals.

A challenge with transformation goals however in addition to their mutual incompatibility are different levels of importance allocated to transformation goals. Decision-making concerning process standardization thus requires multiple criteria decision-making, which requires to weigh these criteria in advance. Thus, decision-making concerning process transformation goals requires the DSS to weigh goals according to importance in advance to give one goal priority over another via a priority ranking among standardization goals [31]. We formulate MR2c as follows:

MR2c: The DSS needs an importance ranking among goals to select among alternative standard process designs.

Further, decision-making requires both forms of process knowledge to complement each other to overcome mutual weaknesses. As a direct requirement of MR1 and MR2, both types of process knowledge need to be combined in a single comprehensive as-is process model before decision-making. As bottom-up process models from MR1 and top-down process models in MR2 each deliver an incomplete analysis of processes in isolation, both models need to be merged. Thus, we derive MR3:

MR3: The DSS needs combine both bottom-up and top-down process information in a comprehensive as-is process model for decision-making.

In addition to these status quo-oriented meta-requirements to derive a comprehensive As-Is process model, an additional meta-requirement is established concerning the possible future process state against which the as-is process model is to be matched to derive a future standard process recommendation. To select the most suitable standard process, the DSS needs to possess a repository of potential standard specifications concerning the future target process design from which an optimal process design in X' can be chosen. We formulate MR4 accordingly:

MR4: The DSS needs a repository of best-practice standard process models.

Furthermore, the purpose of the DSS is to provide decision support between the process model alternatives in the form of process standardization. One method to compare process models with standard best-practice models of ERP systems is the application of business process similarity [22] as motivated in the conceptual foundations in Sect. 2. Thus, the proposed DSS relies on business process similarity to minimize the distance between the input models and the target models [32] in the repository (MR4). We formulate the final MR5 accordingly:

MR5: The DSS needs a similarity-based matching logic to propose an appropriate future standard process design.

Following the introduction of the meta-requirements, the next section will introduce design principles to conceptualize the DSS. Section 6 contains the description of the implementation in the Apromore process analytics platform [12].

5 Design Principles and Design Decisions

We translate the meta-requirements into design principles (DPs) and design decisions (DDs) to steer the later development of the software artifact and to modularize the components of the DSS. According to MR1, the DSS is required to incorporate bottom-up process information. In turn, this requires to extract relevant process data from information systems and prepare the information for process mining in an event log database. Further, the event log needs to be visualized in a graphical process model such as a BPMN representation. Thus, DP1 is formulated as follows:

DP1: The DSS provides a bottom-up process mining layer to retrieve business processes and associated information from organizational information systems.

To account for DP1, we a data extraction program needs to be implemented in the information systems to extract the relevant process mining data (DD1.1). Further, the raw data needs to be transformed into a process mining event log (DD1.2). Finally, the process mining event log needs to be visualized graphically in a process model formalization such as BPMN to be able to unify both bottom-up and top-down information (MR3) and to perform the attribute-based similarity matching of the as-is model against the to-be standard process models. Thus, the DSS includes a BPMN visualization engine (DD1.3).

Further, MR2 requires the DSS to incorporate de jure process knowledge into decision-making, which requires the provision of a user interface to enrich the bottom-up process mining models with additional top-down information which can otherwise not be retrieved by process mining such as additional shadow-process steps or intangible contextual process attributes outside of information systems. We formulate DP2 as:

DP2: The DSS provides the ability to enter additional top-down information.

To identify the contextual information which needs to be attached, we consulted literature on business process standardization to identify relevant process standardization attributes. In particular, the contribution by Romero et al. [33] retrieves a collection of contextual factors which impact the extent of process standardization. In their contribution, the authors find the extent of standardization to be determined by six process categories, namely process activities, resources, data, control-flow, information technology, and management. For each of these categories of contextual factors, we retrieved several sub-attributes from literature which can be assigned with either a numeric attribute value or string of characters for matching (DD2a). Thus, process models need to be attached with these top-down process standardization attributes to perform later similarity matching of the as-is process against the possible to-be standard processes. We formulate DP2a accordingly:

DP2a: The DSS provides process standardization attributes as one element of top-down information.

As standardization attributes refer to different aspects of processes such as the entire process, a specific process variant, or to the task-level, attributes need to be added to the respective level accordingly.

Furthermore, MR2b demands the incorporation of process transformation goals. Therefore, we formulate DP2b to require the DSS to provide a list of possible transformation goals. DP2a is expressed accordingly:

DP2a: The DSS provides process transformation goals as one element of top-down information.

To translate DP2a into a design decision, we performed a series of workshops with the six senior managers responsible for process transformation at the industry partner in the SAP S/4 HANA migration project to retrieve a collection of process transformation goals. Results for process transformation goals in addition to standardization include flexibility, efficiency, cost reductions, compliance, integration, process stability, transparency, measurability, simplification and complexity reductions, and sustainability, which will be given as possible matching values (DD2b).

In addition, MR2c requires the possibility to specify a relative importance prioritization to these standardization goals. Hence, DP2c demands a prioritization of the standardization goals and attributes:

DP2c: The DSS provides a priority ranking for process standardization attributes and standardization goals as one element of top-down information.

To account for DP2c, the DSS allows to weigh each attribute and transformation goal with an importance factor between 0 and 1 to adjust the relative weight assigned to the respective element in the similarity matching algorithm (DD2c).

In addition to the incorporation of bottom-up (MR1) and top-down (MR2) process information, MR3 requires to combine both types of process knowledge before decision-making in the algorithm to determine the most suited standard process.

DP3: The DSS needs to combine bottom-up process mining models and associated top-down information in an enriched process model of the as-is process.

The proposed DSS accounts for DP3 with a visualization module which combines bottom-up process mining models with standardization attributes in an enriched BPMN 2.0 model of the as-is process (X) (DD3).

To be able to propose a suited standard process specification, the enriched as-is process model needs to be matched against the different possible process designs as required by MR4. To implement the requirement, DP4 is formulated accordingly:

DP4: The DSS needs access to a repository of different best-practice standard processes designs.

To be able to perform the attribute-based similarity matching algorithm which uses the extended BPMN model of the as-is process as input, the standard process models in the repository need to be in the same format and be attached with additional top-down information. Thus, the proposed DSS contains a repository of BPMN 2.0 process models (X_S'), such that the as-is process can be matched against each of the process models in the repository to determine the models with a high degree of similarity as candidate for standardization (DD4).

Finally, the last requirement MR5 refers to the need of a matching algorithm which determines the similarity of the as-is process (X) for each of the candidate process models (X_S') in the standard process repository to recommend a target model for implementation. We formulate DP5 as follows:

DP5: The DSS needs a similarity-based matching algorithm for matching of the enriched as-is process model against best-practice standard models in the repository.

Recently, "process similarity" has gained a high degree of attention and numerous approaches to process matching have been proposed. By means of a literature review, several potential process matching techniques were identified and compared to select attribute-based similarity matching as a suited candidate to solve the problem at hand. The contribution by Becker and Laue [34] categorizes process similarity measures into approaches including the correspondence between process model nodes and edges, the edit distance between graphs, causal dependencies between the different activities, and similarity approaches based on trace sets. For example, the contribution by Dijkman et al. [35] identifies five similarity dimensions to be taken into account, namely syntactic, semantic, attribute-based, type-based and contextual similarity. Therefore, the authors propose to measure the similarity from three aspects including node-matching, structural, and behavioral similarity. Finally, Thaler et al. [36] introduce natural language, graph structure, behavior, and human estimation as determinants of model similarity.

Most of similarity matching techniques are based on the model structure or behavior and define distance metrics between a pair of process models to quantify the similarity. The authors in Li et al. [37] provide an approach to measure the structural similarity between business processes based on the number of transformation operations such as adding, deleting or moving to change the structure from one business process to the other. A frequent challenge in process matching are differing labeling styles between process models. For example, a verb-object label like "create order" refers semantically to the same task as the action-noun style "creation of order". To address the issue, the algorithm relies on natural language processing. Thus, the "BPMNDiffViz" by Ivanov et al. [38] compares process models in BPMN 2.0 language using label matching and structural matching metrics. The ICoP Framework by Weidlich et al. uses structural similarity to identify matches and correspondences between business processes [39]. In sum, the calculation of process model similarity needs to take into account heterogeneity of behavioral representation, labeling styles and terminology [40], as well as process model structure [35]. However, for the proposed DSS, the measurement of similarity needs to be extended to take into account process model attributes such as the attached standardization information. Thus, standard process recommendations are derived through an attribute-based similarity matching algorithm which calculates process model similarity for each variant of the as-is process model against the to-be standard process models in the repository based on process model attributes, behavior, structure, and text processing of labels (DD5). Figure 1 summarizes the conceptualization of the different modules of the DSS based on the meta-requirements and design principles.

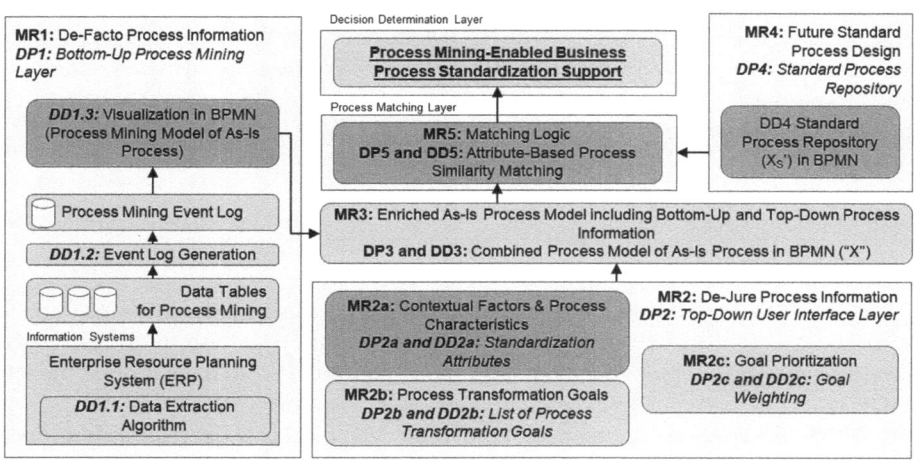

Fig. 1. Conceptualization of the decision support system and modules (Blue: Essential) (Color figure online)

6 Implementation in the Apromore Platform

We associate a number of benefits to the implementation in Apromore, which is an open-source collaborative online business process analytics platform provided by the Apromore Initiative [12]. Specifically, with regard to both the wide acceptance in the community and the rich functionalities provided, we decided to implement the DSS in Apromore. In addition to the workshops performed in the SAP S/4 HANA project context at the industry partner to enrich the meta-requirements from academia with practical insights, the Apromore DSS uses real-world data from three SAP R/3 ERP systems from three sub-companies of the manufacturing corporation.

To account for DP1, we implemented a data extraction program in each of the SAP R/3 systems of the industry partner to extract the relevant data tables required for process mining as .csv-files (DD1.1). Further, the raw data in individual .csv files needs to be translated into a process mining event log. Thus, our solution imports all relevant data into an SQL database to perform the event log generation by a SQL transformation script. To perform the event log generation, a German process mining company provided the transformation script for the purpose of this research to generate the event log from the SAP raw data (DD1.2) in the SQL database. Finally, we export relevant information from the event log into .xes-files for Apromore (Table 1).

Table 1. Overview over process mining event log

Process	Purchasing			Sales		
Company	A	B	C	A	B	C
Period	01.01.2016–31.07.2017					
Number of cases	998,80 Thsd.	432,21 Thsd.	108,54 Thsd.	15,8 Mil.	65.377	155.125
Number of process variants	20,67 Thsd.	10,47 Thsd.	2,54 Thsd.	35,32 Thsd.	39,815 Thsd.	20,87 Thsd.
Total number of process steps [Millions]	4,13	2,15	0,34774.	106,52	50,49	6.07
Avg. number of process steps	4,13	4,98	4,42	6,74	6,02	8,37
Distinct process steps	30	154	54	21	21	22

In principle, the DSS implementation in Apromore can be adapted to incorporate other and any forms of process mining event logs. Finally, we use the BPMN visualization functionality provided by the Apromore platform [12] to draw process models (DD1.3).

Further, the Apromore instantiation provides a graphical user interface as illustrated in Fig. 2. The user interface allows to attach standardization attributes which are valid for either the entire process, a particular process variant, or a specific task (DD2a). Further, project information such as transformation goals can be entered through a list of possible transformation goals (DD2b) and be prioritized through a numeric weight factor between 0 and 1 (DD2c).

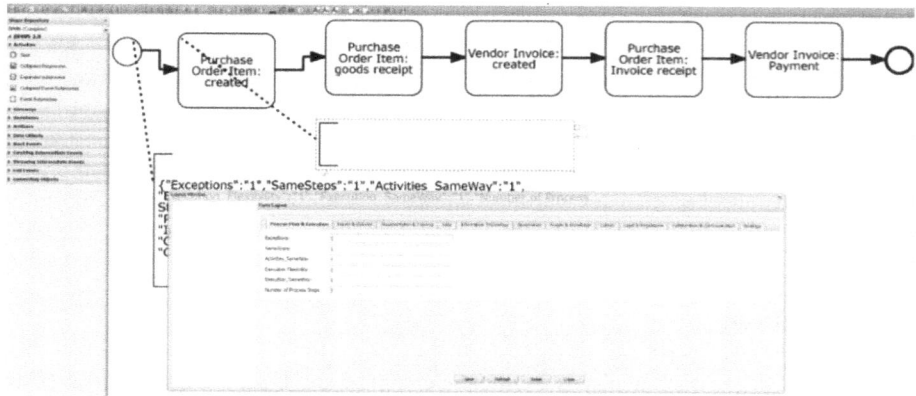

Fig. 2. Graphical user interface to attach top-down process information

Furthermore, the information needs to be combined in an enriched BPMN process model of the as-is process (X) according to (DD3). We use BPMN annotations for visualization in Apromore to display the additional top-down process information.

To retrieve the repository of standard business processes according to DD4, we downloaded the library of standard process specifications from the SAP SolutionManager and imported the library into Apromore as matching candidates. In addition to other ERP management functionalities, SAP SolutionManager is a comprehensive tool to perform business process management and documentation for the SAP ERP landscape. SAP SolutionManager 7.2 provides a publicly available database of to-be standard processes in BPMN 2.0 language for SAP systems. Each of the to-be process models was enriched with the standardization attributes and assigned with values in a workshop with 6 process experts to be able to implement the algorithm.

Finally, to perform the attribute-based similarity matching according to DD5 under consideration of the additional standardization attributes and process transformation goals, we developed a new similarity-based matching plugin based on the existing "similarity search" plugin in Apromore. The algorithm performs matching in three steps. The first-level matcher performs matching at the process-level to ensure the as-is process is matched against the correct domain of the to-be processes such as sales or procurement processes in the repository and considers process-level standardization attributes. Further, each variant of the as-is process differs from the other variants in terms of graph structure, variant behavior, and standardization attributes. Thus, the second variant-level matcher calculates the similarity of each variant of the as-is process according to behavior, graph structure of the variant, and the difference between attribute values. Third, the task-level matches similarity of tasks and attributes. Compared to existing approaches, process models do not contain additional top-down information such as process standardization attributes, which requires the algorithm to consider similarity of attached standardization attributes. For each top-down attribute, the numeric distance is computed. Distances are multiplied by attribute weights and divided by the number of attributes to achieve a weighted similarity score of an individual task within the variant. The overall similarity for a to-be process in the repository is calculated by the sum of variant similarities weighted by the number of variant occurrences. The final result of the attribute-based similarity-matching algorithm in the DSS is thus a similarity measure between 0 and 1 (1 = perfect similarity) for each of the to-be standard processes in the repository. Thus, decision-makers receive a list of all standard processes ordered by descending similarity to the as-is process. The algorithm displays the final similarity score report for each of the to-be standard process models in the repository ordered by descending similarity.

7   Evaluation

Evaluation of the artifact quality is a critical element of any DSR project (e.g., [29]). To determine the ability of the DSS artifact to achieve the intended purpose, three aspects of the DSS are evaluated separately in the first design cycle. First, feasibility of the DSS is evaluated by applying the DSS for the purchasing and the sales process in company A. Second, process decision-makers are asked for their experiences on the DSS in semi-structured interviews. Third, visualization forms of the attribute-extended process models are evaluated in terms of process comprehension. In the second design cycle, an evaluation of the process matching algorithm will be conducted by means of comparison against the matching performed by human users.

In the technical feasibility evaluation, we considered the number of variants to cover a threshold of at least 80% of cases for each process. For the purchasing process of company A, 41 variants were taken into account which cover a span of 869,63 thousand purchase orders and assigned with the standardization attributes on the process-, variant-, and task-level in a workshop with three purchasing process experts. After application of the similarity matching algorithm, the proposed target standard process was the standard end-to-end procurement process from SAP which achieved the highest similarity score of 0,87. Likewise, for the sales process of the company, 56 variants were processed to cover 12,74 million sales orders. As the as-is process contains a large number of customer-specific adaptations, the algorithm produced a comparably low degree of similarity of 0,68 for the SAP standard process specification "Sales from Stock Direct Sales" for the new S/4 HANA ERP system. Table 2 presents results for the application of the DSS for the purchase-to-pay and the order-to-cash processes for one sub-company of the manufacturing corporation.

When asked for their opinion on the DSS and the helpfulness in process standardization, managers highlighted the ability of the DSS to support the selection of a suitable standard process and to justify the decision due to the reliance on data from the ERP system from process mining. Further, managers liked the DSS as it allows for analyses of the required changes to the process before the implementation of the new standard process. Managers further stated the DSS further helped them in advancing BPM as a core capability of the organization, and to increase the "process-oriented thinking" of their employees and themselves. However, managers further highlighted the effort to attach all top-down standardization information to the process variants, and the requirement to implement process mining in a pre-project.

Table 2. DSS Results for Purchasing and Sales Process of Company A

Process	Purchase-to-Pay ("Purchasing")	Order-to-Cash ("Sales")
Company	A	A
Number of cases considered	869,63 Thsd.	12,74 Mil.
Number of variants considered	41	56
Number of different tasks	30	15
Similarity score of proposed standard process	0,87	0,68
Proposed target standard	SAP_E2E_P2P Standard_Procurement	SAP_E2E_O2C Sales_from_stock_Direct_Sales

Second, four different forms of representation of the extended process models were evaluated in terms of process model comprehension to determine how well process decision-makers are supported in their understanding of processes and the associated standardization information. A structured literature review was undertaken to identify impact factors on process model comprehension. Based on the results, three influencing factors on comprehension, namely visualization, decomposition and visual guidance were selected for the design of enriched process models which differ in their representation of the standardization attributes. The first representation lists the standardization attributes in an additional table next to the BPMN process model. The second visualization statically integrates the attributes directly into the BPMN process model and links the attributes in branches to the entire process, the variant, or an individual task, respectively. The third visualization form copies the second form, but additionally provides an interactive component which lets users dynamically show and hide the attributes. The fourth process model realizes visual guidance features and illustrates attributes and values with icons. These four process models were evaluated by an online experiment with n = 8 process experts and n = 35 students regarding their impact on the dependent variable process model comprehension in terms of the time required to answer comprehension questions to the process models and the standardization attributes. Comprehension was operationalized by effectiveness (the number of correct answers), efficiency (time spent on answering) and relative efficiency (proportion of effectiveness through efficiency) as well as by the subjective measurement of perceived ease of comprehension. Results indicate that the static visualization form without the dynamic interactive feature achieves the highest process model comprehension in terms of efficiency and relative efficiency. Besides, the fourth visualization form with visual guidance has the highest effectiveness and is perceived as the easiest form to comprehend. Regarding subjective preferences of respondents, respondents preferred the guided process model with icons (n = 20 respondents who ranked the variant as their highest preference; 46,51%) and the static process models extended with branches (n = 12; 27,91%), while the interactive process model extended with branches (n = 4; 9.30%) and the process model extended with a table (n = 1; 2,33%) achieved the lowest result. Thus, the second design cycle will implement the guided process model with icons.

8   Conclusion and Outlook

Although state-of-the-art information systems increasingly provide organizations with tremendous amounts of process data, and process mining delivers mature techniques to turn data into process information, turning information into actual process decisions remains a substantial challenge. This paper designs a process mining-enabled DSS to aid organizations in process standardization. By extending "de-facto" process models from process mining with additional "de jure" process information in decision-making, the DSS might considerably improve the ability to standardize business processes. However, the DSS also encounters several limitations and requirements to the second design cycle. First, the DSS determines the process model with the highest degree of similarity from the repository of best-practice standard processes. Although "similarity" implies a minimization of organizational change and thus lower tangible and intangible costs for implementation, the "best" candidate for implementation might be a more radical change towards a process with only a low degree of similarity to the as-is process. Second, to match business process against models in the repository, the to-be standard models need to be attached with top-down information, which might differ between organizations and thus not generalize to other contexts.

References

1. Teece, D.J.: Business models, business strategy and innovation. Long Range Plan. **43**(2–3), 172–194 (2010)
2. Tiwari, A., Turner, C.J., Majeed, B.: A review of business process mining: state-of-the-art and future trends. Bus. Process Manag. J. **14**(1), 5–22 (2008)
3. van der Aalst, W.M.P., Weijters, A.J.M.M.: Process mining: a research agenda. Process/Workflow Min. **53**(3), 231–244 (2004)
4. van der Aalst, W.M.P.: Process Mining: Discovery, Conformance and Enhancement of Business Processes. Springer, New York (2011). https://doi.org/10.1007/978-3-642-19345-3
5. van der Aalst, W.M.P.: Process mining in the large: a tutorial. In: Zimányi, E. (ed.) Business Intelligence, vol. 172, pp. 33–76. Springer, Cham (2014)
6. Indulska, M., Green, P., Recker, J., Rosemann, M.: Business process modeling: perceived benefits. In: Laender, A.H.F., Castano, S., Dayal, U., Casati, F., de Oliveira, J.P.M. (eds.) ER 2009. LNCS, vol. 5829, pp. 458–471. Springer, Heidelberg (2009). https://doi.org/10.1007/978-3-642-04840-1_34
7. van der Aalst, W.M.P.: "Mine your own business": using process mining to turn big data into real value. In: Proceedings of the 21st European Conference on Information Systems (2013)
8. Loebbecke, C., Picot, A.: Reflections on societal and business model transformation arising from digitization and big data analytics: a research agenda. J. Strateg. Inf. Syst. **24**(3), 149–157 (2015)
9. van der Aalst, W.M.P., Reijers, H.A., Weijters, A.J.M.M., et al.: Business process mining: an industrial application. Inf. Syst. **32**(5), 713–732 (2007)
10. Wurm, B., Schmiedel, T., Mendling, J. et al.: Development of a measurement scale for business process standardization. In: ECIS 2018 Research-In-Progress Papers (2018, forthcoming)

11. Botta-Genoulaz, V., Millet, P.A., Grabot, B.: A survey on the recent research literature on ERP systems. Comput. Ind. **56**(6), 510–522 (2005)
12. The Apromore Initiative (2018). Apromore: Advanced Process Analytics Platform. http://apromore.org/about
13. Donaldson, L.: The Contingency Theory of Organizations. Sage, Thousand Oaks (2001)
14. Sousa, R., Voss, C.A.: Contingency research in operations management practices. J. Oper. Manage. **26**(6), 697–713 (2008)
15. Der Aalst, V., Wil, M.P., Dustdar, S.: Process mining put into context. IEEE Internet Comput. **16**(1), 82–86 (2012)
16. Melão, N., Pidd, M.: A conceptual framework for understanding business processes and business process modelling. Inf. Syst. J. **10**(2), 105–129 (2000)
17. Vom Brocke, J., Zelt, S., Schmiedel, T.: On the role of context in business process management. Int. J. Inf. Manage. **36**(3), 486–495 (2016)
18. Škrinjar, R., Trkman, P.: Increasing process orientation with business process management: critical practices'. Int. J. Inf. Manage. **33**(1), 48–60 (2013)
19. Galbraith, J.R.: Designing Complex Organizations. Addison-Wesley, Reading (1973)
20. Haußmann, C., Dwivedi, Y.K., Venkitachalam, K., et al.: A summary and review of Galbraith's organizational information processing theory. In: Dwivedi, Y.K., Wade, M.R., Schneberger, S.L. (eds.) Information Systems Theory, vol. 29, pp. 71–93. Springer, New York (2012)
21. Sprague, R.H.: A framework for the development of decision support systems. Manage. Inf. Syst. Q. **4**(4), 1 (1980)
22. Fischer, M., Heim, D., Janiesch, C., et al.: Assessing process fit in ERP implementation projects: a methodological approach. In: International Conference on Design Science Research in Information Systems 2017, Karlsruhe, Germany, 30 May–1 June 2017
23. Harmon, P.: Business Process Change: A Guide for Business Managers and BPM and Six Sigma Professionals. Morgan Kaufmann, San Francisco (2010)
24. Nissinboim, N., Naveh, E.: Process standardization and error reduction: a revisit from a choice approach. Saf. Sci. **103**, 43–50 (2018)
25. Schönig, S., Cabanillas, C., Jablonski, S., et al.: A framework for efficiently mining the organisational perspective of business processes. Decis. Support Syst. **89**, 87–97 (2016)
26. Agrawal, R., Gunopulos, D., Leymann, F.: Mining process models from workflow logs. Computer Science. IBM, San José (1998)
27. Markus, M.L.: Technochange management: using IT to drive organizational change. J. Inf. Technol. **19**(1), 4–20 (2004)
28. Hevner, A.R., March, S.T., Park, J., et al.: Design science in information systems research. Manage. Inf. Syst. Q. **28**(1), 75–105 (2004)
29. Venable, J., Pries-Heje, J., Baskerville, R.: FEDS: a framework for evaluation in design science research. Eur. J. Inf. Syst. **25**(1), 77–89 (2016)
30. Rosemann, M., Vessey, I.: Toward improving the relevance of information systems research to practice: the role of applicability checks. Manage. Inf. Syst. Q. **32**(1), 1–22 (2008)
31. Afflerbach, P., Leonhard, F.: Customer experience versus process efficiency: towards an analytical framework about ambidextrous BPM: completed research paper. In: Thirty Seventh International Conference on Information Systems, Dublin (2016)
32. Martens, A., Fettke, P., Loos, P.: A genetic algorithm for the inductive derivation of reference models using minimal graph-edit distance applied to real-world business process data. In: Tagungsband Multikonferenz Wirtschaftsinformatik, pp. 1613–1626 (2014)
33. Romero, H.L., Dijkman, R.M., Grefen, P.W.P.J., et al.: Factors that determine the extent of business process standardization and the subsequent effect on business performance. Bus. Inf. Syst. Eng. **57**(4), 261–270 (2015)

34. Becker, M., Laue, R.: A comparative survey of business process similarity measures. Process/Workflow Min. **63**(2), 148–167 (2012)
35. Dijkman, R.M., Dumas, M., van Dongen, B.F., et al.: Similarity of business process models: metrics and evaluation. Inf. Syst. **36**(2), 498–516 (2011)
36. Thaler, T., Schoknecht, A., Fettke, P., et al.: A comparative analysis of business process model similarity measures. In: BPM 2016 Workshops, pp. 310–322 (2016)
37. Li, C., Reichert, M., Wombacher, A.: On measuring process model similarity based on high-level change operations. In: Li, Q., Spaccapietra, S., Yu, E., Olivé, A. (eds.) ER 2008. LNCS, vol. 5231, pp. 248–264. Springer, Heidelberg (2008). https://doi.org/10.1007/978-3-540-87877-3_19
38. Ivanov, S.Y., Kalenkova, A.A., van der Aalst, W.M.P.: BPMNDiffViz: a tool for BPMN models comparison. Business Process Management Conference 2015 Demo Papers, Innsbruck, Austria, 31 August–3 September 2015
39. Weidlich, M., Dijkman, R.M., Mendling, J.: The ICoP framework: identification of correspondences between process models. In: International Conference on Advanced Information Systems Engineering, pp. 483–498 (2010)
40. Dijkman, R.M., van Dongen, B.F., Dumas, M., et al.: A short survey on process model similarity. In: Bubenko, J., Krogstie, J., Pastor, O., Pernici, B., Rolland, C., Sølvberg, A. (eds.) Seminal Contributions to Information Systems Engineering: 25 Years of CAiSE, pp. 421–427 (2013)

Author Index